Engineering Ethics in Southern Africa

Theories and Cases

Engineering Ethics in Southern Africa

Theories and Cases

By: Lorrainne Doherty

Engineering Ethics in Southern Africa: Theories and Cases

First published 2023

Juta and Company (Pty) Ltd
First Floor, Sunclare Building, 21 Dreyer Street, Claremont 7708
PO Box 14373, Lansdowne 7779, Cape Town, South Africa
www.juta.co.za

ISBN 978 1 48513 285 1 (Print)
ISBN 978 1 48513 286 8 (WebPDF)

Project Specialist: Fuzlin Toffar
Editor: Rod Prodgers
Proofreader: Lee-Ann Ashcroft
Cover Designer: Renaissance Studio
Typesetter: LT Design Worx
Indexer: Language Mechanics

Typeset in ITC Stone Informal Std 10.5pt on 14.5pt

TABLE OF CONTENTS

1: Moral Philosophy

2: Professionalism

3: Engineering Case Studies

Case Study: The Kariba Dam Collapse 145

Case Study: The Merrispruit Dam Disaster.................................... 151

CONCLUSION

KEY TO THE ICONS

The following icons have been used throughout to highlight important text. Here are some examples of the icons in use.

Quote

Ethics should not be merely an analysis of ordinary mediocre conduct, it should be a hypothesis about good conduct and …

Key Roles

1: REGISTRY

ECSA has the role of providing a *registry* for engineers. The ECSA registry provides for both candidate engineers and professional engineers.

Code of Conduct

The ECSA Code of Conduct

Like many other professions, ECSA's Code of Conduct for Registered Persons (https://www.ecsa.co.za/regulation/RegulationDocs/Code_of Conduct.pdf) …

Case Study

CASE STUDY: THE MERRIESPRUIT DAM DISASTER

In 1978 Harmony Gold Mine designed and built Merriespruit Dam No 4 at Harmony Gold Mine in Virginia in the Free State in South Africa (https://www.tailings.info/casestudies/merriespruit.htm). It was a *'tailings dam'* and was built within 300 m of the first house.

PREFACE

Iris Murdoch (2001: 76) famously wrote the following in her paper 'The Sovereignty of Good Over Other Concepts':

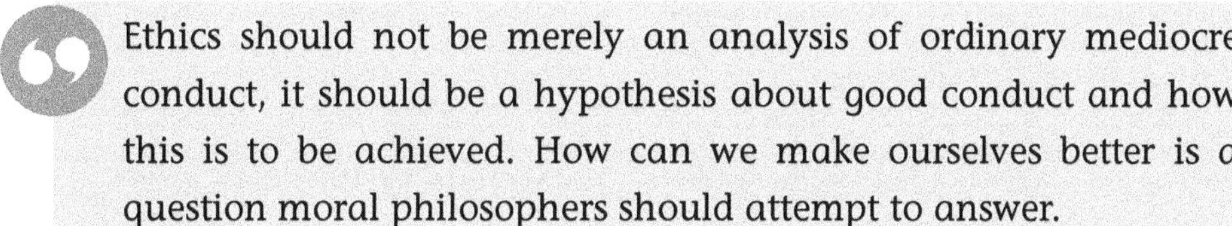

> Ethics should not be merely an analysis of ordinary mediocre conduct, it should be a hypothesis about good conduct and how this is to be achieved. How can we make ourselves better is a question moral philosophers should attempt to answer.

This book, *Engineering Ethics in Southern Africa: Theories and Cases,* the first of its kind for South Africa, attempts to show within its pages how indeed engineers can 'make themselves better' in terms of their ethics in both their personal and their professional lives. In doing so, they will need to heed the words of Archibald Alexander (1914: 16), who endorses the view that;

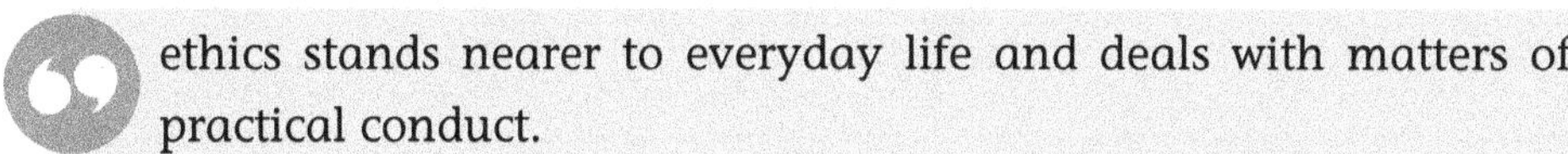

> ethics stands nearer to everyday life and deals with matters of practical conduct.

Alexander's observation is particularly pertinent for contemporary engineers, whose profession increasingly places them and their work in an 'everyday' social context.

Ethics is undoubtedly practical and is considered an applied and interdisciplinary discipline, drawing on both science and the humanities that in turn enable it to provide the tools for improving our understanding of, amongst other things, what is 'right'; what is 'wrong'; what is 'good'; what is 'bad'; what is 'fair'; what is 'unfair'. Ethics also serves to address our expectations of both the meaning and reality of these terms in our personal and professional lives, the latter as engineers invariably extending into the wider community. Furthermore, the aim of ethics is to guide us as individuals and the wider community to a state of 'well-being' and/or 'flourishing' with its implicit promise of happiness and all that it entails.

As an academic tradition, the study of ethics (moral philosophy) spans in excess of 2 000 years and its importance has grown with the ascendance of science as we seek to understand how we should employ our moral reasoning, moral decision-making, moral conduct and moral action when confronted with situations of moral dilemma and/or moral conflict in an everchanging and increasingly unpredictable world. The contents of this book endorse that each one of us has moral agency which we can exercise to ensure that both our own ethics and morals and those of wider society are authentic and practical.

In conclusion, through theory, practice and case studies, this book attempts to show contemporary engineers what being ethical means and not only *'how they may be better'* as professionals, but also *'how they should know better'* too.

References

Alexander, A (1914) *Christianity and Ethics. A Handbook of Christian Ethics.* London: Duckworth & Company.

Murdoch, I (2001) *The Sovereignty of Good Over Other Concepts.* US: Routledge Classics.

ABOUT THE AUTHOR

Dr Lorrainne Doherty (BSc Econ Hons; MPhil; DPhil) lectures at the University of the Witwatersrand in the School of Mechanical, Industrial and Aeronautical Engineering. Since 2013, she has become a regional expert in Engineering Ethics.

Prior to her return to academia in 2010, she had a successful career in local and international marketing. Brands under management included Trivial Pursuit®, Star Wars®, Five Roses®, Willards®, Johnson Wax Rally® and Pennzoil®.

Her research interests lie in the practical application of philosophy, ethics and organisational behaviour.

She guest lectures to industry.

DEDICATION

When approached by Juta to author this book, I was excited to learn that it would be the first book of its kind dealing specifically with Engineering Ethics in South Africa. I have spent the past ten years at the School of Mechanical, Industrial and Aeronautical Engineering at the University of the Witwatersrand teaching this subject and wanted to share the knowledge I had gained with both students of engineering and those currently practising engineering. It would be fair to say that my initial thought was the book was best viewed as my 'legacy project'. That said, it has taken on a life of its own and would not have been possible to complete without the support and guidance of some very special people.

First, I would like to thank my colleagues, Professor Robert Reid, Professor Craig Law and Doctor Bernadette Sunjka, who have encouraged and supported me every step of the way in what has become an all-consuming project. Their assistance has proved invaluable and I am deeply indebted to each of them for their understanding.

I would also like to extend my heartfelt thanks to my family, who have understood the scope of this task and what it means to me. Their love, understanding and faith in me, has provided me with confidence when at times I doubted my ability to complete this project in such a short time-span. My absence at many get-togethers with friends and family has been accepted in good grace by them and I feel sure that once the book has been published, a celebration with them all is the very least I can arrange to convey my love for them and sincere gratitude for their constancy.

My thanks also need to be expressed to the staff at Juta & Company (Pty) Ltd, and none more than Melissa Toerien, who has been my 'go-to' lady and guide. We have travelled a road together, unfamiliar to myself, and she has provided me with all the signposts I needed. Your help and trust in me has been much appreciated, Melissa.

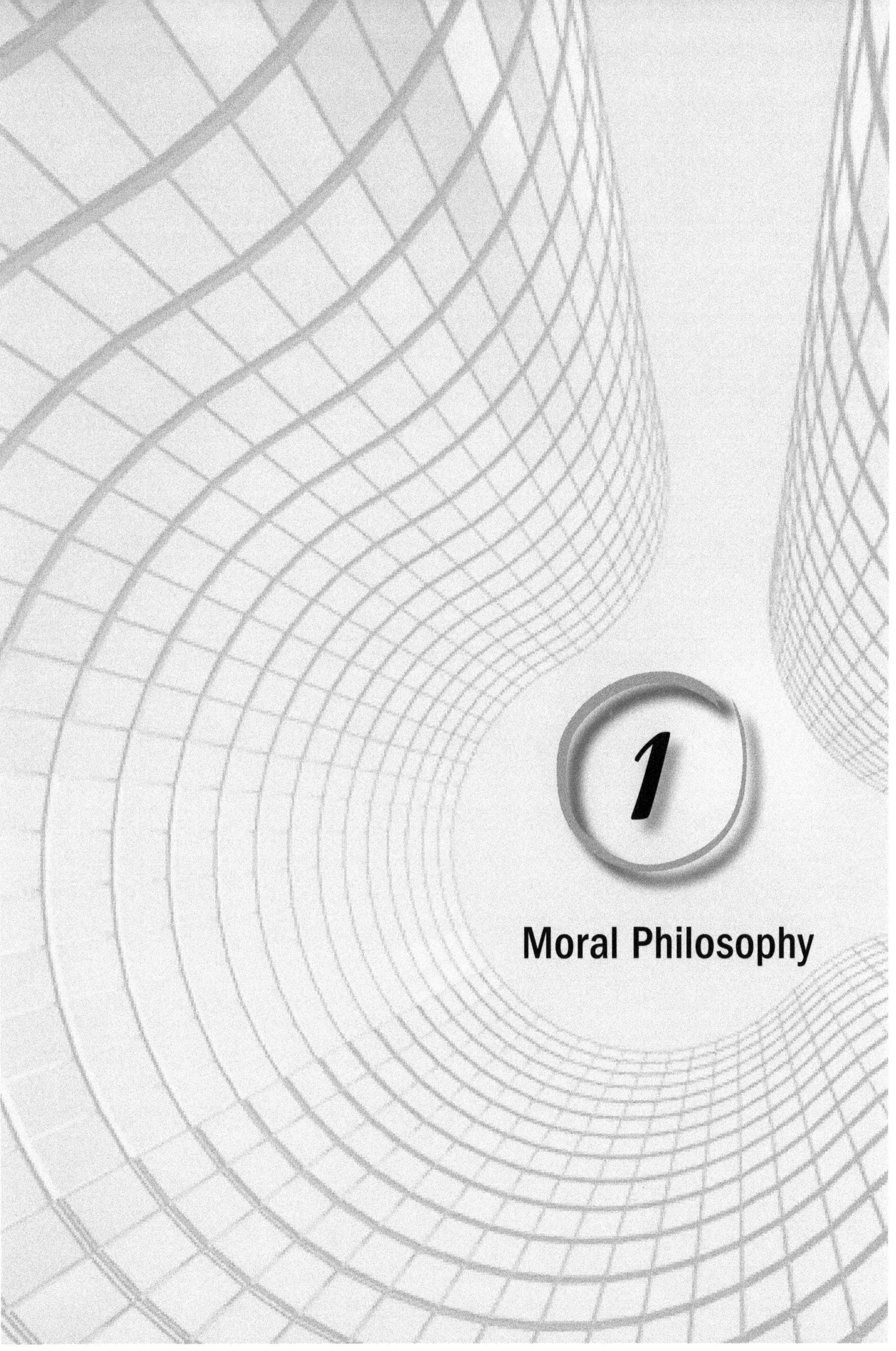

1

Moral Philosophy

A BRIEF INTRODUCTION TO MORAL PHILOSOPHY

The etymology of the word *'philosophy'* can be traced back to the Greek word *'philos'*, meaning love, and the Greek word *'sophia'*, meaning using one's intelligence in practical affairs. When seen together in the word philosophy, they are most usually interpreted as the 'love of wisdom'. Philosophy has been defined by *The Concise Oxford English Dictionary* (1995) as;

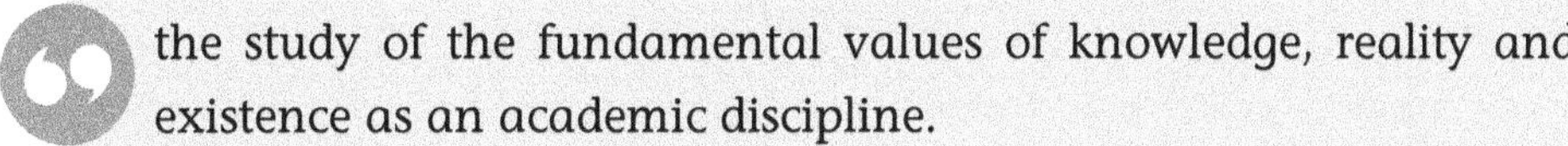

> the study of the fundamental values of knowledge, reality and existence as an academic discipline.

The Oxford Dictionary of Philosophy (2008: 275) defines philosophy as;

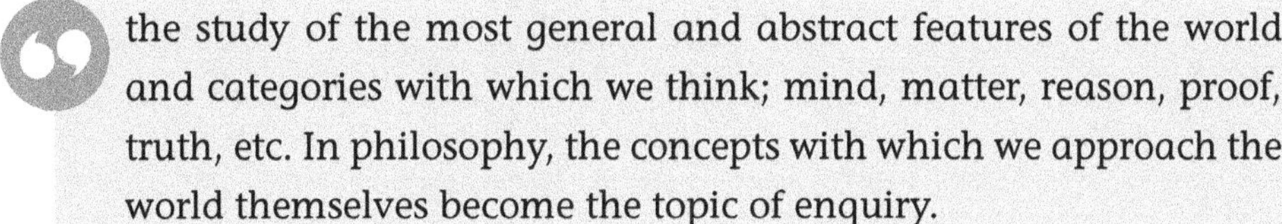

> the study of the most general and abstract features of the world and categories with which we think; mind, matter, reason, proof, truth, etc. In philosophy, the concepts with which we approach the world themselves become the topic of enquiry.

As an academic discipline, philosophy is both rigorous and disciplined and is rooted in words, meaning and debate that asks difficult questions of us such as: 'What is truth?' 'Is there such a reality as moral universalism?' 'Is the contemporary understanding of human dignity rooted in economic value?' 'Does the concept and meaning of freedom have inherent limitations?'

In trying to answer such questions and in order to adopt suitable academic positions, philosophers construct rational arguments for their views and in so doing demonstrate they follow and embrace a logic from more basic principles. In so doing, philosophers construct arguments opposed to those of their opponents that will move the debate forward, which in turn help us to understand the strengths and weaknesses of the different positions taken (or posed) and the relationships between them. This method used by philosophers can be referred to as the three Cs of philosophy (see Figure 1.1) and, when employed, it reveals how philosophy provides an extremely useful antidote to dogmatism as it serves to highlight issues and draws distinctions that we need to make in order to advance our thinking about a subject, topic or our position on either.

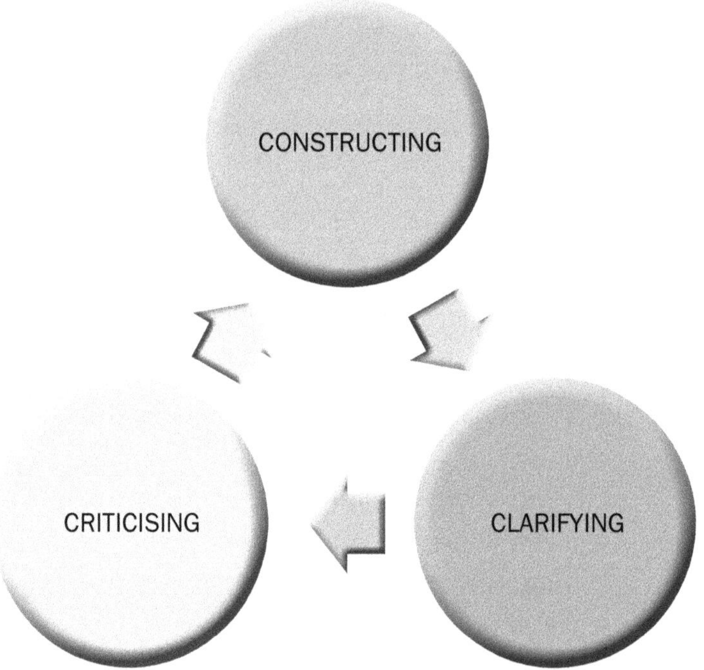

Figure 1.1: The three Cs of philosophy.
Source: Author's own.

Of importance in advancing the benefits of the three Cs is the notion that words have a precise meaning. In the same way that numbers provide us with precision, answers and certainty, for example $2 + 2 = 4$ (it can never be 3 or 5), words can provide us with the same certainty if used correctly and sparingly. Words therefore are the tools of philosophy and philosophers the craftsmen who use these tools to their best advantage. When used correctly, words provide a coherence to views expressed and a defence against criticism. They also serve to avoid confusion and argument, vagueness and assumptions.

Within the discipline of philosophy, writing is seen to be an extension of reading, discussion and debate. To this end, systematic argumentation is required not just within the context of debate, but also within the context of academic writing in this very distinct, academic tradition.

PHILOSOPHY THROUGH THE AGES

It is generally accepted there are distinct eras accompanying Western philosophical thought (see Figure 1.2). These eras correspond directly with recognised approaches that are associated with specific philosophers, some of whom we study later in this book. These eras in philosophy are commonly accepted as falling into four key periods, namely: the Ancient or Classical; the Medieval; the Modern; and the Contemporary. The Ancient or Classical era is rooted in Greek philosophy in which not only were the tradition and discipline of philosophy graced by some extraordinarily talented minds, but these minds extended their knowledge to include moral philosophy or ethics as we now know it. The Medieval era was a time when Christianity was a major influence in the study and tradition of Western philosophy with some of the best-known philosophers being directly associated with the Church. Their philosophy addressed and directly impacted the lives and minds both of the faithful and those of dissenters too. The Modern era coincided with the Age of Enlightenment and the ascendance of science. Steeped in rationality, this era coincided with new scientific discoveries and technologies. Meanwhile, the Contemporary era is associated with the behavioural aspects of philosophy and its impact upon both individuals and their relationships, their environment and each other. New ethical approaches such as animal rights, feminism, care ethics, environmentalism and identity ethics are just some of the approaches that have gathered academic traction and scholarship during this era.

It has been suggested by some academics that we are currently moving away from the Contemporary era in philosophy towards a new era that recognises the importance of the Fourth Industrial Revolution (4IR) and the cyber technologies, robotics and computerisation that accompany it. In recognition of this, moral philosophy has been required to further scrutinise and challenge our existing understanding of both metaphysics and meta-ethics and has developed approaches that address and respond to contemporary moral issues, dilemmas and conflicts the like of which are addressed by ethical approaches such as techno-ethics, bio-ethics, machine ethics, transhumanism, and privacy and freedom. When developing these 'new' approaches that arise due to 'new' scientific discoveries and 'new' technologies, philosophers have required a 'new' focus that has been developed in conjunction with 'new' terminology to explain the 'new' moral challenges, conflicts and dilemmas that are occurring.

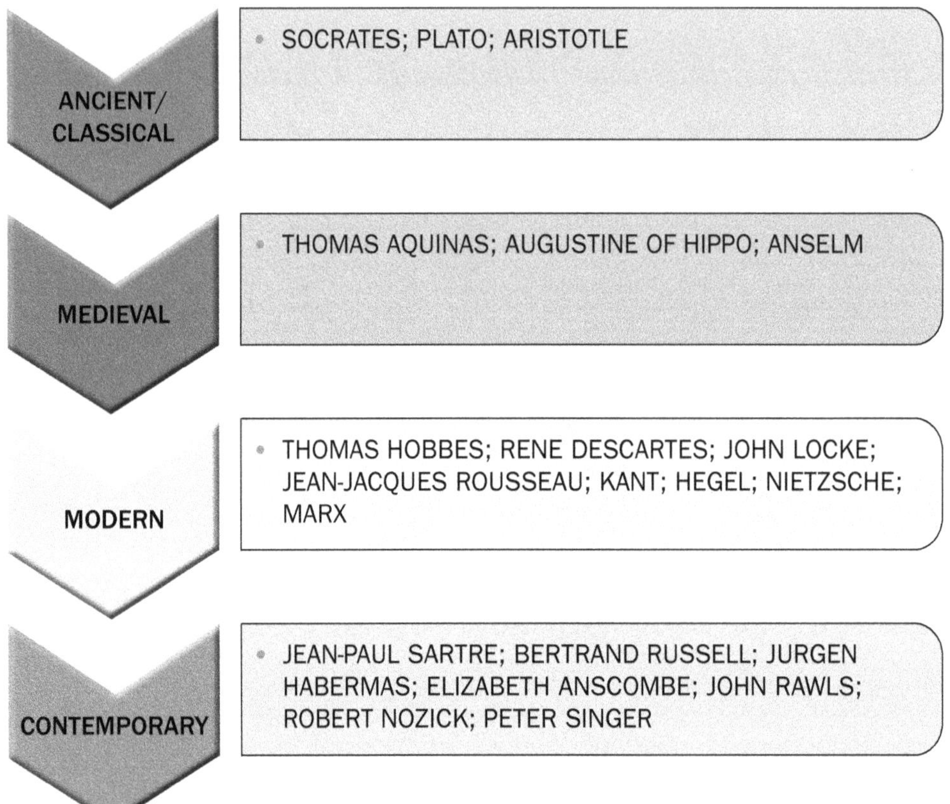

Figure 1.2: Accepted eras within Western philosophy.
Source: Author's own.

WHERE DOES ETHICS FIT IN?

Within the Ancient/Classical era, about 2 000 years ago, Greek philosophy developed a separate branch within the discipline called Moral Philosophy. It continues to play an important role in the philosophical tradition as it encompasses the tradition of *'moral reason'*. Arguably, it is best interpreted as applied philosophy and is used in our everyday personal and professional lives, often instinctively, but also within a learned capacity. By reading this book, engineers should have the opportunity of examining their personal ethics in addition to their professional ethics and in so doing provide themselves with the further opportunity of examining the ethical expectations of themselves, their profession and the wider community.

ETHICS AND MORALS: ARE THEY THE SAME?

In practice, these two terms are used interchangeably. This hints at the relationship of interdependence these two terms enjoy in society, in the workplace and for the individual. Together they contribute to and provide us with a direct sense of:

- Order

- Harmony

- Expectations

However, in philosophical theory the terms 'ethics' and 'morals' are specific, distinct and different. Each serves a precise purpose with its own set of consequences. At this stage we have perhaps uncovered the need to turn to academic definitions to assist in our interpretation and understanding of the two terms as it will serve to provide us with an explanation for their interdependent relationship.

DIFFERENCES BETWEEN 'ETHICS' AND 'MORALS'
Ethics

Ethics = Values/Beliefs

The word 'ethics' comes from the Greek word *'ethikos'* and relates to ethos or character. It concerns human judgements such as:

- Good/Bad

- Praise/Blame

- Fair/Unfair

- Better/Worse

- Right/Wrong

Our judgements are rooted in our values and beliefs and thus can be considered **relative** and/or **conditional** because what one person judges to be *'good'* another person might judge to be *'bad'*. Similarly, one person might judge an individual worthy of praise because they have met the passmark for an assessment, whereas another would consider this achievement not praiseworthy because in meeting the minimum requirement for passing the assessment, the candidate has done so with the bare minimum of marks. This example serves to remind us of the importance that values play in our lives. It is proposed they are not only deeper than attitudes and more embedded in our character, but perhaps even longer lasting. Our values are developed over time (the course of our lifetime from birth until our death) and are subject to relative forces (see Figure 1.3 below).

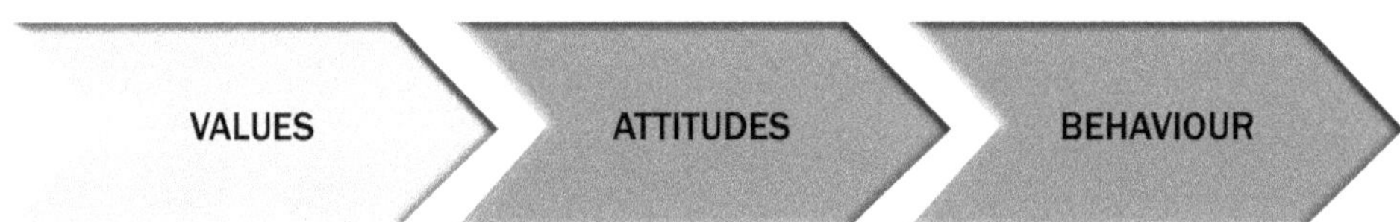

Figure 1.3: The importance of values.
Source: Author's own.

In Figure 1.3 we see that values are at the root of human experience. The suggestion is that our values shape our attitudes, which in turn inform our behaviour. Of importance is that no one is value-free – values are learned. Not one of us is born a racist, misogynist or feminist. These values are learned over time and become reflected in the attitudes and behaviour of those we believe hold them. A good example of this can be seen in South Africa post-democracy. The policy of apartheid was racist and developed personal values rooted in the superiority of one race over another. This arguably shaped the attitudes of both white South Africans and South Africans of colour that in turn informed the behaviour of all South Africans during this time in history. By contrast, Nelson Mandela in our 'new' South African democracy urged every South African to recognise and adopt the vision of being an important part of the *'rainbow nation'*, which many argue provided a 'new' national value suitable for a non-racial democracy with unity and harmony as the basis that would inform the behaviour of all in the country.

However, it is perhaps to Epictetus the Stoic we must turn if we are to fully understand the importance of values and the way in which they shape our attitudes, which in turn informs our behaviour. The following words and sentiment have been ascribed to him:

 Men [*sic*] are disturbed not by things, but the view which they take of them.

Morals

Morals = Norms/Accepted Practices

The words 'moral' and 'morality' come from the Latin *'moralis'*, a concept concerned with actions, not issues of human character. It arguably speaks directly to our behaviour. It concerns issues of:

* Just/Unjust

* Punishment/Clemency

* Guilt/Innocence

* Acceptable/Unacceptable

Unlike ethics, it concerns accepted norms and practices and thus can be considered absolute/unconditional but, most importantly, universal. Within society, there are invariably accepted 'ways of doing things'. Sometimes these are codified through laws, whilst sometimes they are just accepted practices, such as giving up one's seat on a bus or train to an elderly and/or infirm person. In any event these norms can arguably be interpreted as representing societal values, which speaks to the interdependency of ethics and morals and how they can influence one another. If we view this within a South African context, we can identify with and understand why South African society protects the vulnerable, punishes those who are found to have committed acts of gender-based violence (GBV) and has committed to the rural development programme (RDP) in which government homes are provided for those who were previously disadvantaged and are without permanent shelter.

THE GOALS OF SUCCESSFUL ETHICS AND MORALS

Now that we have explored the definitions and realities of ethics and morals and their interdependence, it is perhaps important to understand what the goals of each are. Are they different or do they have a common purpose? It is argued the latter position is the key to their independent and mutual success. Their goals are best summed up as being three-fold. There is a need for:

- Authenticity = Morals are organic. They adapt over time and/or as situations demand and have legitimacy for both the individual and the wider society.

- Relevance = Morals need to provide the recognised means to address an issue.

- Practicality = Morals and their consequences need to be easily implemented.

All three aspects cited above are required if morals are to be the active means to guide both the individual and wider society towards the goals of **order**, **harmony** and **expectations** when situations of moral dilemma and/or conflict occur.

WHAT SHAPES ETHICS AND MORALS?

We first need to recognise that ethics and morals are not stagnant. They are organic and can be shaped by our relationships, environment, experiences and aspirations. To fully understand the importance of this we could perhaps ask ourselves the following question: 'Do my ethics and morals coincide with those of my parents, friends, associates or political party?' We also need to acknowledge the role that extrinsic forces play in developing and shaping our ethics and morals. **Extrinsic forces** are best understood as being *outside forces that have a direct bearing on outcomes*.

However, whilst we have a working definition for extrinsic forces, we need to understand how they present themselves and affect us in our daily life, specifically within the context of our ethics and morals. Listed below are examples of some extrinsic forces that can shape and change our personal values and those of wider society.

Extrinsic forces

- Economics

- Politics

- History

- Culture/Tradition

- Religion

- Law/Rules

- Media (modern media includes social media, journalism, entertainment)

To demonstrate the 'direct bearing' these forces have had on 'outcomes', I have listed below some of the significant and very real shifts and changes that have occurred in our moral and ethical outlooks and actions that have been directly influenced by the presence and impact of extrinsic forces:

- Slavery

- Apartheid

- Homosexuality

- Child labour

- Animal experimentation

- Abortion

- Basic education

- The sale of contraceptives and the day-after pill

- Universal suffrage

WHAT IS ETHICAL/MORAL?

After briefly examining the etymology, the definitions and the tradition of ethics and morals, it is important that we turn to developing an understanding of their reality in our everyday lives. We can do this by asking ourselves the questions 'What is ethical?' and 'What is moral?' These seemingly simple questions have a reality that is often much more complex and it is not uncommon for an air of personal confusion to be present as we examine our stance in situations that present themselves as moral dilemmas and/or moral conflicts.

As irrational as it seems, it is often by knowing and/or understanding what is **unethical** that we get a sense and understanding of what is **ethical**.

We can do this by applying our skills of observation, knowledge and understanding in a way that helps us reach a decision on whether it impedes or prohibits either our own or the broader community's **flourishing/well-being**. If we use this distinction, it can be found to be extremely useful in assisting us when we consider issues where blurring can easily occur and our decisions on ethicality and morality seem confused and lack substance and consensus. The following photographs graphically illustrate situations where flourishing/well-being has been impeded and there is a distinct lack of order, harmony and expectations, which in turn would point to the unethical in South Africa today.

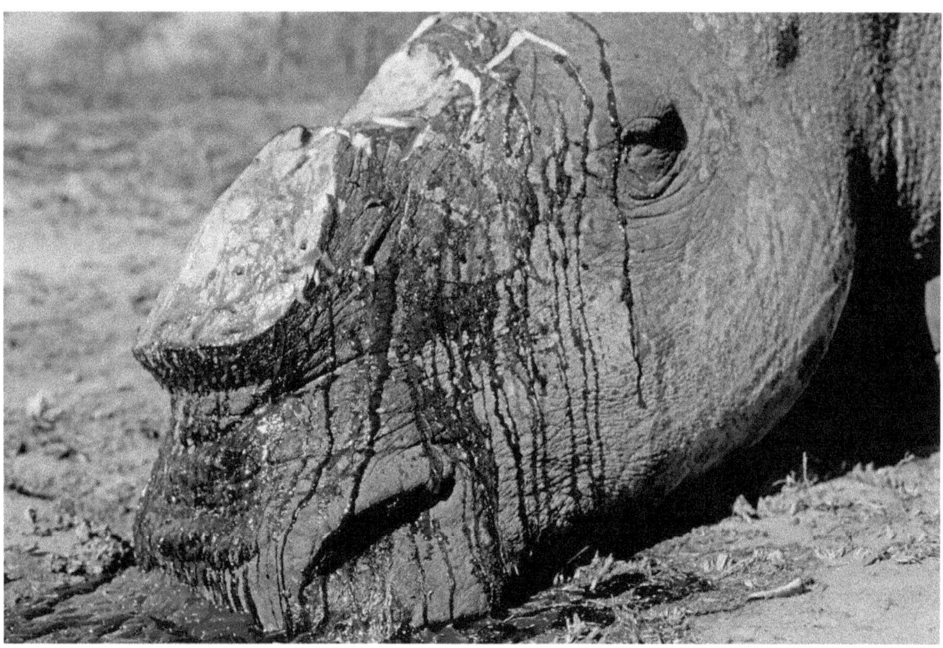

Figure 1.4: Rhino poaching.
Source: https://www.gettyimages.com/photos/rhino-poaching

Figure 1.5: Xenophobia in South Africa.
Source: https://www.gettyimages.com/photos/xenophobic-attacks-in-south-africa

THE IMPORTANCE OF ETHICS AND MORALS IN CONTEMPORARY SOCIETY

Contemporary philosophers are in unison when it comes to the importance of moral philosophy (ethics and morals). I would like to share some of their thoughts in order that we can further understand not just the importance of this applied philosophy in our everyday lives and within our work as engineers, but also its critical importance in and for South Africa and the world around us.

Morality is ultimately practical: though it matters morally what we think and feel, morality is, at its heart, about what we do.

– Kwame Appiah (2011)

> There has been a shift of emphasis in philosophical discussion of ethics, away from the purely abstract questions to more practical ones . . . This shift of concern towards 'applied ethics' has been beneficial.
>
> – Jonathon Glover (2001)

> To be interested in ethics is to be interested in life! . . . Faced with two equally convincing arguments, how do you decide which is 'right'?
>
> – Mel Thomson (2010)

> Ethics is central to modern life.
>
> – Vardy & Grosch (1999)

From these scholarly observations of the role that ethics and morals plays in our life, it is perhaps to Geschwindt (2006: 316) that we should turn for the final say on the importance and reality of ethics for us all in our personal capacity and as professional engineers:

> If the world is to become a more peaceful place, it is vital that ethical decision-making becomes an integral part of life for us all. The ability personally to make the right ethical decision in difficult circumstances and under pressure, taking account of all concerned, is a necessary ingredient for a good life – a flourishing, decent life, for whomever, wherever, whenever.

References

Appiah, K (2011) *The Honor Code: How Moral Revolutions Happen.* US: WW Norton.

Blackburn, S (2008) *The Oxford Dictionary of Philosophy.* Second Edition Revised. Oxford: Oxford University Press.

Geschwindt, S (2006) *Am I Right? Or Am I Right? An Introduction To Ethical Decision Making.* Canada: Trafford Publishing.

Glover, J (2001) *Humanity. A Moral History of the Twentieth Century.* US: Yale University Press.

The Concise Oxford Dictionary (1995) Ninth Edition. Oxford: Clarendon Press.

Thomson, M (2010) *Understand Ethics.* Oxford: Hodder Headline.

Vardy, P & Grosch, P (1999) *The Puzzle of Ethics.* New Edition. UK: Fount Paperbacks.

LEARNING OUTCOMES

- What are the 3Cs of philosophy and why are they important?

- Explain the difference between ethics and morals, including why their relationship is one of interdependence.

- Name five situations and/or practices which you consider to be unethical.

- Name an extrinsic force and how it has changed an existing value(s) of yours.

- What are the three goals (for both an individual and society) that ethics and morals provide?

THE VIRTUE APPROACH

BACKGROUND TO CLASSICAL ARISTOTELIAN VIRTUE ETHICS

Virtue ethics was developed over 2 000 years in the Classical era and the Greek philosophers Plato and Aristotle are most closely associated with developing the framework and elements of this ethics approach. However, it is to Aristotle that we must turn if we want to understand the full scope and rigour of this theory, which is judged to be normative, human-centric, consequential and ultimately practical. Aristotle, ever the pragmatist, wanted to develop a 'working theory' as experience taught him that a moral life could be achieved by the efforts of the individual (self) that flowed through to the wider community (other) through the adoption of a virtue approach. In her support, Annas (2013: 677) unequivocally states the virtue ethics approach provides us with;

an organised and systematic way of telling us what is the 'right' thing to do.

Aristotle (2019) built on the theoretical and academic work of his predecessor Plato and developed this practical approach to ethics as a self-help guide for men whose aim was to become statesmen in their *'demos'*. A statesman was a social position in Ancient Greek society that was not only considered to be ultimately praiseworthy but also carried significant social recognition, respect and social status. To achieve this political position and the state of *'eudaimonia'*, variously translated as 'well-being' or ultimate happiness, Aristotle proposed that political candidates needed to acquire 'good' character traits that would serve them not just as individuals (self) in their life but could also serve them in assisting the wider community (other) when they became statesmen and held political office, and were required to discharge civic duties and promulgate and enact civic laws.

Aristotle called these all-important character traits *'arete'* (commonly translated as personal excellence) and suggested their acquisition, cultivation and demonstration was a vital part of an individual's **'lived experience'** if they were to become the best versions of themselves. Not surprisingly therefore, Classical Aristotelian virtue theory proposes the result of an individual being *'virtuous'* is they undertake and demonstrate the *'right'* and the *'good'* in their daily lives through their actions and behaviour, as opposed to the 'wrong' and the 'bad'. In short Aristotelian virtue ethics can best be summed up as an ethical approach that encourages us to reach our fullest potential as human beings and recognises that in order to achieve this, we are a **'constant work in progress'** due to the ongoing and persistent need to acquire virtue as our lives take us through different phases and experiences.

Of importance is this ancient, character-based ethics approach has retained its relevance over time (unlike some ethical approaches) and indeed philosophers such as Alasdair MacIntyre (1990), Peter Baron (2014) and Nick Bommarito (2018) have ensured that our current understanding of how modern virtue demonstrates itself in contemporary life in terms of the individual and the community is clear and the consequential benefits meaningful if **'well-being'** and **'ultimate happiness'** is to be attained and maintained. However, what must never be forgotten is that measurability, balance and the impact they have on harmony both personal and collective are axiomatic in the Classical Aristotelian approach to virtue theory. Aristotle proposes that balance and harmony are fundamental if we are to succeed in creating both a moral individual (self) and, by extension, a moral society (other). The importance of these terms is quite clear when examining the Aristotelian theoretical measure of the **'Golden Mean'**.

We must also not forget the binary nature Aristotle attributed to his virtue theory. He saw virtue as the opposite character trait to vice and attested that we are all more than equipped to identify those of us who possess and demonstrate either. Whilst both the words 'virtue' and 'vice' have entered into our everyday vocabulary, their meaning and importance is only truly understood if we examine the approach and theoretical framework of Classical Aristotelian virtue theory, after which hopefully we will enthusiastically embrace the one, whilst wholeheartedly eschewing the other.

KEY DETAILS OF CLASSICAL ARISTOTELIAN VIRTUE ETHICS

The Aristotelian approach to virtue is deeply indebted to the Classical Greek understanding of the triune nature of the human soul, which extended to reason, emotions and appetites, with reason being uniquely human. To this end, Aristotle incorporated this understanding into his dualistic virtue framework, which in essence speaks to reason (intellectual virtue) and emotions (moral virtue) whilst acknowledging that appetites can be a driving force in both. Figure 2.1 provides a simplistic view of the differences between intellectual and moral virtue as understood through the writings of Aristotle.

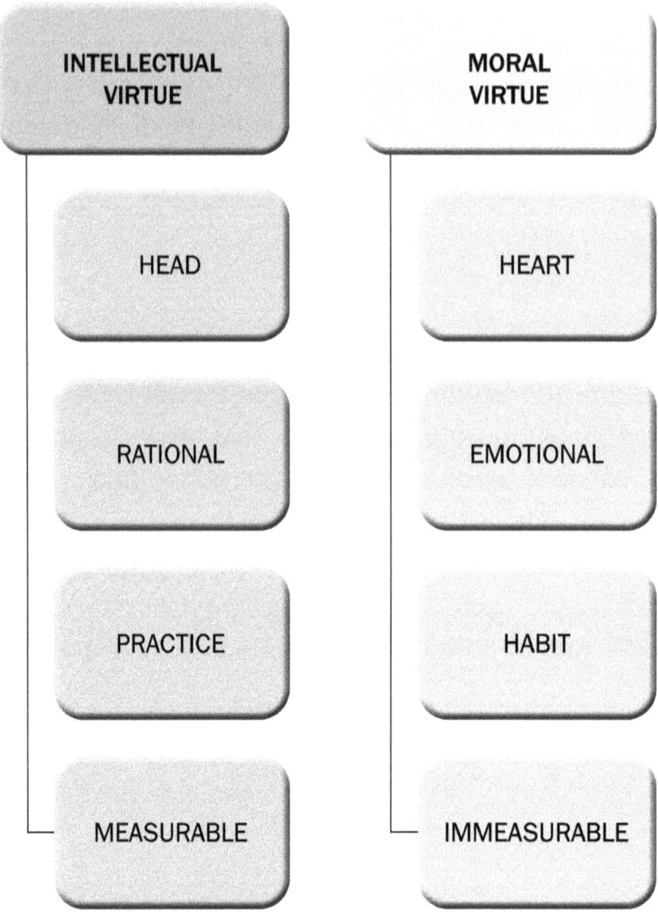

Figure 2.1: Key points of difference between moral and intellectual virtue.
Source: Author's own.

THE CLASSICAL ARISTOTELIAN VIRTUE THEORY FRAMEWORK UNWRAPPED

Importantly, Aristotle hypothesised that you are not born virtuous, but rather you *'acquire'* virtue through *'mentorship'*. Ever the moral realist, Aristotle recognised that an individual observes and tries to consistently emulate their role model or mentor. It is quite natural for one to assume their 'happiness' and 'well-being' can be yours too if you just . . . copy them! These role models or mentors do not need to have celebrity status, suggested Aristotle; indeed in contemporary times they can be members of one's own family, peer group, sports team or political party, or a teacher, religious instructor or one or more work colleagues.

Aristotle proposed the success we attribute to these role models/mentors is due to the virtue(s) they in turn have acquired and demonstrated in their daily lives through their behaviour and actions. Their acquired intellectual and moral virtue(s) arguably provide them with the 'well-being' necessary to assist them in being able to successfully navigate any and all of life's myriad challenges, both moral and practical. By actively demonstrating their virtues (being virtuous) they are seen, according to Garvey and Stangroom (2013: 99), to be;

doing the right thing at the right time with the right feelings in the right way and for the right reasons.

INTELLECTUAL VIRTUE

Classical Aristotelian virtue theory proposes four intellectual virtues (based in the rational and cognitive) namely:

- Resourcefulness

- Technical skill (techné knowledge)

- Judgement

- Scientific knowledge (epistemé knowledge)

These can still be seen to have contemporary relevance and I would suggest more importantly they are still measurable and can directly impact the 'good' and 'bad' and the 'right' and 'wrong' when practised by the individual (self) and the community (other). In short, these intellectual virtues are observable and their consequences measurable.

This is best illustrated (and understood) if we were to examine the situation of an individual being stranded on a desert island with a fresh water supply, yet no means of transport to leave the island. In this scenario, we can see how all four Aristotelian intellectual virtues could provide the castaway with a sense of 'well-being' and 'happiness' when practised, until help arrives and they are able to leave the island with their rescuers.

1. Their *judgement* would assess their current situation, which they would realise was life-threatening without adequate shelter and food.

2. Their ***resourcefulness*** would lead them to the conclusion that a shelter needed to be constructed by themselves to protect them from the elements and that catching fish would be the simplest and safest means of providing food, as it would take the least energy to catch, yet provide sufficient nutrition for their survival.

3. Their ***technical skill*** would ensure they made a shelter that would withstand the elements and a rod and a line suitable for catching the smallest and largest fish they could reasonably land.

4. Lastly, their ***scientific knowledge*** would enable them to site their shelter in a spot that would ensure its construction would remain erect despite weather conditions and also plot the tides to ensure they fished at the correct time and in the place where they would be most likely to succeed in catching fish.

This example demonstrates the underlying rationality and cognitive skills required in engaging these intellectual virtues and their concomitant behaviour and actions in addition to their measurability. Supporting this position, Baron (2014) proposes that intellectual virtues are invariably the result of education and practice, both of which contain the all-important element of measurability, for example what qualifications do you hold; what experience do you have; how many time times have you undertaken the task?

Moral virtue

Many of the Aristotelian original moral virtues do not have resonance in contemporary times and also lack relevance. For example, three of the moral virtues cited by Aristotle were:

- Liberality

- Magnificence

- Wittiness

None of these would be considered moral virtues in modern times, not even for somebody who was planning to be a statesman! Why should we need to acquire liberality as a moral virtue, let alone magnificence or wittiness? Fortunately, some of the other moral virtues that Aristotle identified have retained their importance and relevance despite the passage of time and can still resonate with us. Aristotle theorised that phronesis or 'practical wisdom' was central to his virtue theory, especially when assessing the need and efficacy of any moral virtue. However, I think we can agree the following three moral virtues cited by Aristotle are not only recognised as important in contemporary times but are directed and used in large part by our personal, practical wisdom. They are:

- Patience

- Justice

- Truthfulness

We can all attest to the relevance of these moral virtues as a source of moral agency for both an individual and the wider community in which they live. However, whilst these moral virtues might resonate with us in contemporary times, we would find it difficult, if not impossible, if asked to measure these virtues. How does one measure patience, and how much patience is sufficient for it to be considered a virtue? The same difficulties are associated with measuring justice and truthfulness. How much justice needs to be dispensed for it to be considered enough or too little, and how do we measure if somebody is telling the truth? Surely one is considered to be either truthful or a liar, with no 'degrees' of truthfulness being recognised?

Aristotle recognised the difficulties in 'measuring' moral virtue and developed a measure called the 'Golden Mean' primarily as an ethical tool that could be used to measure or assess the presence and degree of moral virtue present in an individual through the means of incorporating the concept of *'balance'*.

THE 'GOLDEN MEAN'

Aristotle was above all things an empiricist and observed and recorded the world around him. In developing his virtue theory Aristotle brought the same academic perspective and rigour to his virtue framework. He surmised that if intellectual virtues could be measured, their moral virtue counterparts needed to demonstrate an equivalent facility. This led to him developing the measurement of the 'Golden Mean' as a measure or assessment for moral virtue(s). Essentially this was devised as a sliding scale with a midway point between the two extremes (vices) of excess and deficiency, often graphically portrayed as a seesaw or an old-fashioned set of scales. From this it is evident that 'balance' is a word and concept that helps define Aristotelian virtue theory.

THE ACCURACY OF ARISTOTLE'S GOLDEN MEAN

However, Aristotle recognised the results from this initial measure of a moral virtue to be grossly inaccurate and to overcome this he introduced the aspects of *'feelings'* and *'activity'* into the measurement of the Golden Mean. Whilst still not providing an absolute measure, these additions served to support an aspect of moral realism when applied to human behaviour, actions and experience, and the acquisition and demonstration of moral virtue. In short it provided an extra dimension and accuracy that feelings and activity bring to any measurement/assessment of moral virtue. (See Table 2.1.)

Table 2.1: Aristotle's amended Golden Mean that includes feelings and activity.

FEELING	ACTIVITY	EXCESS	MEAN/VIRTUE	DEFICIENCY
Confidence	Running through enemy lines	Recklessness	COURAGE	Cowardice
Altruism	Giving money	Extravagance	GENEROSITY	Stinginess
Fairness	Punishment	Revenge	JUSTICE	Apathy
Calmness	Restraint	Impulsiveness	PATIENCE	Indecision

CONTEMPORARY VIRTUE ETHICS

Whilst precision has never been associated with the Golden Mean and its results have invariably been assumed to be subjective and relative, it nevertheless provides us with the means to assess the presence of moral virtue in an individual, often in a way considered universal. To this end we have examples when there is general recognition of the virtue(s) that need to be demonstrated. This is the case if a medal is to be awarded to a member of the military, or if civic honours are to be bestowed upon a citizen.

After this brief study of the character-based Classical Aristotelian virtue approach to ethics, it is not an academic leap to see how and why it has transcended time and in so doing retained its contemporary relevance. People don't change, do they? We all arguably strive to do what is *'right'* and to be our *'best'* in situations and demonstrate this through our character traits in our actions and behaviour, especially when faced with adversity and challenges irrespective of when and where we live. In short, whilst history might influence the demonstration, recognition and appreciation of moral and intellectual virtue, the virtue approach has nevertheless withstood the test of time and has shown an ability to be organic and develop and cultivate virtues or character traits (both moral and intellectual) in keeping with the times in which we live.

To this end, an important contemporary concept that is now associated with virtue ethics is that of **intrinsic** and **extrinsic** virtue. The distinction between the two can often serve to provide us with a moral compass not just in our personal lives, but in our professional lives too.

INTRINSIC VIRTUE
Good in and of itself, for example faith, activism, loyalty

EXTRINSIC VIRTUE
Good for the sake of something else, for example accountability, compliance, governance

ARISTOTELIAN THEORY ADAPTED

More recently David Brooks (2016) has examined virtue theory and provided a contemporary context to establish its significance in terms of modern moral agency both in terms of the individual (self) and the community (other). He suggests there are still two types of virtue, that are acquired through self-knowledge, but he has interpreted them as *'Résumé'* virtues and *'Eulogy'* virtues (see Table 2.2). The former would be considered in Aristotelian terms to refer to *'the self'* and the latter to *'the other'*. He proposes that résumé virtues consist of character traits that we actively pursue in our personal and professional lives that contribute to our external success and are driven by self-interest. Eulogy virtues, on the other hand, refer to those character traits by which we wish to be remembered by our family, friends, colleagues and the community and are largely service-driven and the result of what Brooks (2016) terms *'moral logic'*.

Table 2.2: Brooks' modern interpretation of virtue: Résumé and Eulogy virtues.

Résumé virtues (self)	Self-interest, fame, success, status
Eulogy virtues (other)	Kindness, honesty, faithfulness, loyalty

Interestingly, Brooks (2016) does not perceive modern virtue as linear, or as having a capacity of either/or; rather he encourages an individual to develop both types of virtue in order for them to reach their true potential as moral agents and to experience the world in which both 'self' and 'other' are in harmony. In short, he is recognising the essential 'balance' in life to which Aristotle alerted us. This allows us not only to fulfil our true potential as individuals but also encourages us to experience a world in which morality is a touchstone for both an individual (self) and the community (other) that in turn informs a definite and expected sense of 'right' and 'wrong'. In accepting this we are also recognising that our acquisition of virtue serves not just ourselves, but the world in which we live and thus essentially and most importantly speaks to the way in which we would hope to be treated by others.

WEAKNESSES OF THE VIRTUE ETHICS APPROACH

- Aristotelian virtues in their literal sense are no longer applicable, for example magnificence, liberality.

- This approach relies heavily on relativism. What is cowardice? What is justice? The interpretation of terms can change over time as can our understanding of them, for example historically justice was seen to be served only within a retributive context. There was little or no understanding that restorative justice could be appropriate and effective.

- This approach requires mentorship. How can we acquire virtues if we have no example or role model?

- This approach addresses virtue in a solo capacity. Surely there is a case where duty could be incorporated? For example, is there not a duty for a politician to acquire the moral virtues of honesty and loyalty? Or an engineer to acquire the moral virtue of integrity and the intellectual virtue of competence?

- Aristotelian virtue ethics is based on relationality whereas the Fourth Industrial Revolution (4IR) is associated with technology, machines and transhumanism and pays little attention to human relationality.

STRENGTHS OF THE VIRTUE ETHICS APPROACH

- After 2 000 years virtue theory still resonates and has modern relevance arguably because its focus is people in both their capacity as individuals (self) and as members of a community (other).

- The original theory can be extended beyond the individual in contemporary times to incorporate community, corporates and organisations. The latter has gained traction because of the 'personalities' many organisations adopt.

- Virtue (both moral and intellectual) extends to shared/common values and ways of doing things and provides 'meaning', for example armed forces, religion, political parties, schools, families and professions have common values.

- Gender-based virtues have been identified in ethical approaches such as feminism and care ethics.

- It could be argued the concept of 'whistle-blowing' stems from a consciousness and demonstration of moral and intellectual virtue.

- Whilst 4IR is driven by intellectual virtue, there remains a requirement for personal moral virtue, especially within the engineering profession, that can extend to the inanimate in activities such as design, project planning, material integrity, patent and copyright of new techniques and products and the privacy of client information.

VIRTUE ETHICS AND ENGINEERING

Virtue ethics plays a key role in the life of an engineer in South Africa. Not only does it form a vital function in the Engineering Council of South Africa's (ECSA's) professional code of ethical conduct for engineers, but it also informs engineering activities and projects undertaken by engineers within a social context. Therefore, engineers can be judged not just by colleagues, but by their profession and the community for their acquisition and demonstration of virtue. Their engineering activities and materials used can also be seen to possess virtue, for example materials can be judged to have 'integrity', whilst projects can be judged to have 'transparency'.

In their undertaking of all aspects of engineering work, engineers require a moral compass in both their professional and personal capacities and virtue ethics can provide this both in the way in which they use their intellectual virtue and the way in which this is supported by their moral virtue. Moreover, the virtue ethics approach does not expect us to possess virtue immediately; rather it proposes that our acquisition of virtue is to be considered a 'work in progress' as we try and achieve *eudaimonia* for ourselves personally, our profession and the community in which we live and serve as engineers. In line with this it would be fair to assume the goal of any engineering project and the engineers in charge of its design and progress is *eudaimonia*, ie to increase 'well-being' and promote 'happiness' and flourishing.

Engineers can acquire the virtues required by their profession through mentorship by colleagues and engaging in the ethical codes of conduct. To this end, the Washington, Sydney and Dublin Accords subscribe to accepted and universal virtues which they expect accredited engineers to demonstrate both in terms of the way in which they conduct their professional lives and the way in which this assists the profession to which they belong. This virtuous behaviour from engineering professionals is expected by their colleagues, their professional councils, and the wider community whom they serve.

References

Annas, J (2013) 'Being virtuous and doing the right thing' in *Ethical Theory: An Anthology* edited by R Shafer-Landau. Oxford: Wiley-Blackwell, pp 676–85.

Aristotle (2019) *Nichomacean Ethics.* 3rd edition. Translated by Terrence Irwin. Indianapolis, IN: Hackett Publishing.

Baron, P (2014) *Virtue Ethics.* Ethics Study Guide. London: PushMe Press.

Bommarito, N (2018) *Inner Virtue.* Oxford: Oxford University Press.

Brooks, D (2016) *The Road To Character.* London: Penguin Books.

Garvey, J & Stangroom, J (2013) *The Story of Philosophy. A History of Western Thought.* London: Quercus.

MacIntyre, A (1990) *After Virtue.* 3rd edition. US: University of Notre Dame Press.

LEARNING OUTCOMES

- Why and how does virtue ethics differ from other ethical approaches?

- What moral and intellectual virtues do you think a modern engineer needs in their professional life?

- List four vices that have been identified in the contemporary engineering profession.

- Who would you consider to be your mentor? Do you have different mentors from whom you can acquire different virtues?

- Are moral and intellectual virtues of equal importance in the engineering profession?

DEONTOLOGY/KANTIAN ETHICS

Immanuel Kant (1724–1804) was a German philosopher who, we are told, whilst leading a seemingly unremarkable personal life, significantly changed the focus of moral philosophy forever. Such was the magnitude of his influence that his approach has been eponymously named Kantian ethics. Kant took the concept, reality and moral philosophy of duty and crafted a moral paradigm that pushed and continues to push our moral imagination and boundaries.

The word 'deontology' has its roots in the Greek word *deon* meaning 'duty' and this is the thrust of Kantian ethics, which at its simplest can be described as a duty approach to ethics. However, as we shall learn, Kantian ethics incorporates a number of key and important moral philosophical positions within the framework of its theory and maxims, confirming that Kant was originally directing his moral philosophy solely towards the individual. Of interest is that subsequently Kantian ethics has been developed, expanded and extended to include the workplace, organisations and wider society, all of whom we have decided also have a role and function of duty to play in contemporary life and communities.

WHAT IS KANTIAN ETHICS?

Kant sought to develop a universal ethical approach in the absence of theology and metaphysics, which was a significant departure from previous moral philosophical approaches. His philosophy was explained in his two famous works, *The Critique of Pure Reason*, published in 1781, and *The Groundwork of the Metaphysics of Morals*, published in 1785. It is the latter work that contains his framework for his theory of deontology, which has rationality, respect and free will as its foundation and which is further distinguished by its anthropocentricity. However, whilst using rationality or reason as the basis of his duty theory and as the grounds for its anthropocentricity, Kant seemingly ignores children and the mentally challenged who cannot reason. In this regard, whilst Kant considers his moral framework offers a universal moral approach, it is often argued by scholars to be constrained by his neglect of those who cannot reason and becomes the basis for those who argue its theoretical limitations. Critics of the Kantian approach often put forward the view that if directed at human beings, a major flaw in Kantian ethics is that Kant does not recognise the dualism of reason and feelings/emotions in human beings. Kant neglects emotions or 'sensibilities' as having the ability to 'cloud' ethical decision-making rather than enhance it. Mizzoni (2010: 122) highlights this aspect of Kantian ethics when he advises that:

> Some critics though regard Kant's articulation and defense [*sic*] of deontological ethics as an over-intellectualised ethic. They point out that Kant's concept of 'person' as a 'rational being' really describes a pseudo-person, not a real life person who is much more than a rational being.

However, what must not be forgotten is that Kant's unique moral approach nevertheless also acknowledges the duty of respect that should be afforded to oneself and others, which critics argue is not always a function of rationality and free will. Kant theorises the existence of our fellow human beings has an absolute worth in and of itself and by extension the dignity and worth of our fellow human beings is not conditional on any empirical factors. This is a foundation stone of the Kantian theoretical framework which entrenches the purpose and ethical necessity of duty.

Essentially morality for Kant was not just about **'what'** you do; it was critically about **'why'** you do it and it was these two distinct positions that distinguished his duty ethics approach and theory of deontology from any of its predecessors and which continually succeeds in presenting moral and ethical challenges when used by us today.

Kant proposed that we are all independent moral agents because we have the unique human attribute of rationality/reason. Kant proposed that our rationality and reason provide us with freedom from which we can exercise our free will and accept our duty/duties, sometimes also referred to as responsibilities or obligations. He further posited that our moral principles are best understood as being *'a priori'*, or independent of experience, because of our unique ability to reason.

Throughout his philosophical works Kant uses terms that have precise meanings and his Kantian theory was no different (see Table 3.1). When used, Kant assumes the reader will not only understand their theoretical meaning and purpose within the Kantian framework, but recognise their reality in daily life and more specifically within the context of duty, moral agency and his 'new' approach to ethics.

Table 3.1: Key terms in Kantian ethics.

KEY KANTIAN TERM	MEANING
Moral agent	A person with the capacity to act morally
Maxim	A rule/principle
Will	The faculty of deciding, choosing or acting
Imperative	A command

As mentioned earlier, Kantian ethics focuses on the 'why' in terms of ethics and morality and in so doing quite clearly establishes the importance of **intention/motive** derived from a sense of goodwill as the ethical foundation of duty which he sees as a process (see Figure 3.1). Essentially Kant proposes that because humans have free will due to their rationality and consciousness, they are uniquely placed to accept duty(ies).

Figure 3.1: Kantian duty process.
Source: Author's own.

In light of this approach, Kantian philosophy is considered by philosophers to be a **non-consequentialist** approach because Kant determines that it is in fact our intention or motive that is of importance in determining our duty and ergo our ethicality/morality, **not** its consequence. In other words, Kant proposes that it is our intention that causes an action to be *'right'* or *'wrong'*, **not** its consequence. To this end, Kantian theory is proposing that *'the end does not justify the means'*, a position that has caused and still continues to cause significant discourse and dissent amongst philosophers and ethicists alike.

THE DETAILS OF KANTIAN THEORY

When we examine Kantian theory it is important to remember that Kant was extremely aware of the need for human dignity and human rights to provide direction for his theory of duty. This is clear when we take a closer look at Kant's maxims (or rules) that drive the reality of his theory. Kant (2018: 34) writes of the categorical imperative that we must always;

> [a]ct only in accordance with that maxim through which you at the same time can will that it become a universal law.

Interestingly he also arguably adopts a humanist approach (2018: 42) in his theory that advises:

> Act so that you use humanity, as much in your own person as in the person of every other, always at the same time as end and never merely as means.

Once again, we are witness to Kant's determination of human ascendancy and in this particular maxim he has stood accused by some of not providing a 'new' notion for humanity, but rather a secular adaptation of the Christian premise that one should ***do unto others as you would have done unto you'***. In short Kant theorises that we should treat people including ourselves as having intrinsic value, rather than instrumental value. By extension this means that people are valuable in themselves, irrespective of their personal circumstances.

In any event, of importance is that Kant's maxims above are the very foundation of Kant's 'imperatives' (commands) integral to his Kantian approach to ethics, in which duty was front and centre. Essentially, Kant proposes there are two imperatives when a moral agent is called upon to exercise their duty: hypothetical and categorical.

HYPOTHETICAL IMPERATIVE

These duties are contingent or conditional, ie they are based on an **'if'** followed by a **'then'**. Hypothetical imperatives, according to Kant, represent a practical necessity of undertaking a possible action to attain a chosen need. An example of a hypothetical imperative would be, **if** you want to run and complete a marathon, **then** it is your duty to undertake daily training. Another example would be, **if** you do not want to spend time in jail, **then** it is your duty not to actively take part in unlawful activities.

Hypothetical imperatives, as can be seen in these examples of duty, are therefore conditional and do not apply to every individual at all times. They can be context and person specific and have also been described as demonstrating the act of duty in this Kantian imperative as a *'means to an end'*.

CATEGORICAL IMPERATIVE

Categorical imperatives, however, unlike hypothetical imperatives, are considered by Kant to be consistent, impartial, universal and absolute and best interpreted as a binding duty upon all. In other words, the duty should be applied in every case of ethical decision-making irrespective of outcomes. There are no ifs, buts or maybes in Kant's theory of categorical imperative.

Unlike hypothetical imperatives, they are described as being 'ends' in themselves and as such can be seen to be a universal duty or universal moral law. The impact and import of a Kantian categorical imperative or universal duty is simply understood when we use the word **'always'**.

An example of a categorical imperative would be, it is **always** your duty to tell the truth; or another example could be, it is **always** your duty as a member of a national defence force to follow the orders of your commanding officer. If viewed in this way, it helps to clarify the universal and binding nature of this type of Kantian imperative and that it can be a duty that can be applied equally to all. Some would argue this provides an egalitarian stance to the Kantian notion of a categorical imperative in addition to the universality associated with it.

However, we do not need to allow our minds to venture too far into the unknown to grasp the problems associated with the application of a categorical imperative. This is never more apparent than if we were to take a deeper examination of the categorical imperative of **always** telling the truth. Whilst the intention/motive is morally correct, the consequences can be anything but, when applied.

What if we were to apply this absolute and universal duty if we were living in Nazi Germany and harbouring Jewish people in our home in our attic to protect them from the Gestapo and certain death if found. If officers came knocking on our door asking if we were harbouring Jews, would we act upon our categorical imperative of always telling the truth and confirm we were harbouring Jews and watch the Gestapo take them from our home, manhandle them into a lorry bound for a concentration camp and certain death, and be of the opinion that morality had been served through us exercising our categorical imperative and universal duty of telling the truth? In this example, whilst our intention was *'good'*, as truth-telling is one of the most honourable intentions/motives across every society, could we ignore, let alone live with, the outcome/consequence of our intentions as Germans living in Nazi Germany when our duty always to tell the truth translated to sending people to their certain death? I think this example of truth-telling in Nazi Germany within the context of Kantian ethics, and specifically when used as an example of the use of a categorical imperative, highlights the reality of the idiom, *'the road to hell is paved with good intentions'*, and perhaps helps alert us to the challenge of the morality and ethicality of the exercising of a universal and absolute duty.

CONTEMPORARY DEBATE AFFECTING KANTIAN (DUTY) ETHICS

After our brief look at Kantian ethics and the pre-dominance of duty, it is important that we examine the contemporary challenge to this unique paradigm for moral philosophy. Kant assumes the primacy of the individual in terms of duty, born out of our rationality. However, in modern neoliberal welfare societies it is ever more evident the duties of an individual have been increasingly handed over to the State. This handing over of duty or duties to the State, also coincides with the Kantian approach to treat fellow human beings with respect and actively supports the notion of human beings having intrinsic worth. This is especially so when considering vulnerable members of society. To this end, in modern societies we find ourselves asking in our individual capacity, 'Is the State not better equipped and funded to undertake this categorical imperative on behalf of its citizens than an individual?' However, we increasingly see that it is politics that is a driving force in terms of State social interventions and not purely duty. Furthermore, in a world where capitalism is the dominant economic system, we see a fundamental incompatibility with Kantian ethics due to capitalism's profit motive, which pays scant regard to duty or respect as the teleological nature of capitalism justifies and supports monetary ends irrespective of the means used and the intrinsic human need for human dignity.

STRENGTHS OF THE DUTY ETHICS APPROACH

The strengths of duty ethics are six-fold and listed below:

- Duty ethics proposes there are universal standards requiring duties/responsibilities (categorical imperatives) that are to be adopted by all.

- Duty ethics has egalitarianism built in to its philosophical framework as Kant suggests that rationality is the sole purview of humanity.

- Duty ethics does not rely on religious belief. Kant proposes a universal moral standard that exists apart from spirituality/mysticism.

- Duty ethics does not have self-interest at its core. Instead, it proposes respect for oneself and for one's fellow human beings owing to their intrinsic worth.

- Duty ethics is based on reason and rationality. It does not consider emotions/feelings as being integral to ethical decision-making.

- Duty ethics proposes that consequences are not the means by which something is morally 'right' or 'wrong'. Rather the proposed intrinsic in duty ethics is the intention of duty, not outcomes.

WEAKNESSES OF THE DUTY ETHICS APPROACH

The weaknesses of duty ethics can also be seen to be six-fold:

- Duty ethics dispenses totally with the notion of feelings. The theory suggests that feelings serve to cloud and/or blur an issue, rather than inform it. Feelings in Kantian ethics are replaced by a rational sense of duty.

- Duty ethics does not always see the 'bigger picture', especially when the categorical imperative is employed and applied. Intention/motive, whilst seeming morally 'right', can have negative consequences.

- Duty ethics does not provide a means to resolve a conflict of duties, for example as a female, which duty comes first, that of a daughter, a mother, an employee, a friend?

- Duty ethics is anthropocentric and does not recognise other sentient beings.

- The absolutism of duty ethics is problematic when dealing with ethics and morals where relativism has a role to play.

- Duty ethics says that as moral agents, we all have free will. However, how free are we?

KANTIAN (DUTY) ETHICS AND ENGINEERING

Interestingly duty ethics has a significant role to play in contemporary professions including engineering. Duty and virtue ethics invariably form the basis of most of the professional codes of conduct and the same is true for the professional codes of conduct for engineers. This is evidenced particularly in the case of engineers in South Africa in their absolute duty (categorical imperative) to ensure public health and safety at all times.

In common with many professions, because engineers work in a social context, there are accompanying duties that must be undertaken by them when engaging in engineering activities and projects. These are outlined in the professional codes of conduct. Within the context of South Africa, the Engineering Council for South Africa (ECSA) is the source of a professional code of ethical conduct for engineering that it enforces amongst its members.

References

Kant, I (2018) *Groundwork for the Metaphysics of Morals.* Translated by Allen W Wood. UK: Yale University Press.

Mizzoni, J (2010) *Ethics The Basics.* UK: Wiley-Blackwell.

LEARNING OUTCOMES

- Why is Kantian theory unique from its predecessors?

- Why is Kantian ethics considered a non-consequentialist approach?

- Describe the differences between a hypothetical imperative and a categorical imperative.

- Does Kantian theory recognise human feelings and emotions as being instrumental in our ethical decision-making?

- Is human dignity an important component in Kantian ethics and, if so, why?

THE THEORY OF UTILITARIANISM

This ethical tradition has its origins in the eighteenth century. The Scottish philosopher David Hume (1711–1776) is affectionately known as the grandfather of the Utilitarian tradition, but it is the social reformer/ philosopher Jeremy Bentham (1748–1832) whose name is most closely associated with this ethical approach alongside that of John Stuart Mill (1806–1873). The latter, a well-known philosopher, wrote the scholarly work *On Liberty*, before becoming a parliamentarian in the British Parliament, in which he pursued a political agenda that arguably could be interpreted as having utilitarianism as a foundation stone.

Democracy, liberalism, progressive social policies and a pervading sense of optimism were associated with these early Utilitarians who advocated the need for social and legal reform during the first industrial revolution. They theorised this could be achieved if the theoretical *'principle of utility'* or the *'greatest happiness principle'* were applied as a means of moral decision-making, judgement and, most importantly, evaluation in times of moral dilemma or moral conflict. In contemporary times Utilitarianism has a new 'champion' in the Australian moral philosopher and activist Peter Singer (1946–), whose scholarly work can be seen to conform in the main to a Utilitarian approach, although he has adopted the extreme position of Utilitarianism in some of his work which has caused alarm and dissent for both academics and wider society, where it has been argued that Singer has developed some of his ethical positions without recourse to human dignity, arguably a key principle implicit in the Utilitarian tradition.

THE FRAMEWORK OF UTILITARIAN THEORY

Utilitarianism is considered a normative, **consequentialist** theory. To this end, Utilitarianism advises that consequences are what makes something morally 'right' or 'wrong', **not** the intention/motive or the act or behaviour, that might pre-empt and guide it. In short, within this theoretical framework the ends or outcomes, usually perceived as practical results, justify the means.

Mizzoni (2010: 90) explains this quite well when he says:

Utilitarianism is regarded as a consequentialist ethic because the view counsels that in deciding whether an act, rule, policy or motive is morally 'good' we should look to see if it has 'good' consequences for all.

This is a vastly different theoretical framework from that of Kantian ethics, which is non-consequentialist. Table 4.1 graphically illustrates the differences between the non-consequentialist ethical approach of Kant and the consequentialist approach of Utilitarian theory.

Table 4.1: The differences between consequential and non-consequential ethical approaches (Deontology and Utilitarianism).

Deontological/Non-consequential Approach	Utilitarianism/Consequential Approach
Rule-based (imperative) view of ethics was first proposed by Kant and commonly known as Kantian ethics. It is non-consequential as it is intentions and duty based and is not defined by the outcome if considered a categorical imperative.	Consequence-based view of ethics was introduced by Bentham and developed by Mill. It is commonly known as Utilitarianism. Utilitarianism is therefore considered to be ends driven or consequence-based.
'Goodness' or 'badness', 'right' or 'wrong' is determined by the intention.	'Goodness' or 'badness', 'right' or 'wrong' is consequential/teleological and is determined as such by the outcomes (end result).
The only behaviour that can be considered ethical is the one that has duty and goodwill behind it.	Justifies behaviour as ethical if it produces the 'greatest good for the greatest number' and thus produces the greatest utility.

AN EXPLANATION OF THE PRINCIPLES IN UTILITARIANISM

Utilitarianism assesses 'good' by employing the two principles of *'utility'* and *'greatest happiness'* in a theoretical framework that recognises that humans are by nature dualistic as they are both rational and emotional beings. This acknowledged dualism recognises that whilst rationality and reason are an important component of ethical decision-making and morality for humans, equally so are our feelings and emotions. This acceptance of the dualistic nature of an individual represented by the 'head and heart', in the Utilitarian theoretical framework arguably places it as one of the more authentic assessments of moral agency when compared with some of its philosophical counterparts.

In acknowledging the importance of the relative (feelings and emotions), Utilitarianism proposes that *'good'* feelings are associated with *pleasure/ utility*, whereas *'bad'* feelings are always associated with *pain/suffering*. It expands this position further when it suggests that humans have a sense of feelings in common. As such, it creates a sense of *'universalism'* as it proposes that we can all 'feel' and in so doing have a natural sympathy for one another best described by the idiom 'I feel your pain'. The concept and practice of utility also embraces the notion of *'common good'* and is used in conjunction with the moral philosophical concept of flourishing/ well-being. Thus, it also comes with egalitarian overtones as it assumes the 'good' is the same for everyone – what is considered 'good' by me will be considered 'good' by you. In having this understanding, which is central to this theoretical framework, we must remember that utility, whilst based on the premise of pleasure, does not mean the action is painless. Rather it asserts that Utilitarianism embraces the outcome that has the *least pain*. To this end, with these theoretical notions and principles in place, Utilitarianism is often expressed as being an ethical approach that promotes *'The greatest good for the greatest number'*.

HOW IS UTILITY MEASURED?

It has been claimed that Bentham's theory of utility is based on the premise of *'psychological egoism'*. This states that human nature drives us to seek pleasure (happiness) and avoid pain (suffering). However, in Bentham's view, the 'right' or 'good' thing to do is to seek pleasure (happiness) and avoid pain (suffering) and it is this notion that provides our actions with real moral value. Conversely, the 'wrong' or 'bad' thing for us to do is to promote pain, misery and suffering and thereby reduce or minimise any amount of pleasure or 'goodness'.

The understandable question that arises from this stance is how do we measure 'happiness' and 'pain', as Bentham wanted his theory of Utilitarianism and moral enquiry to have the benefits of scientific objectivity. Bentham developed a *'hedonic calculus'* to provide a rational measure for utility, including Bentham's cardinal utility as a means of providing ordinal measurement. To this end, the hedonic calculus provided seven criteria by which to measure the pleasure (happiness) and pain (suffering) produced by any particular action (see Table 4.2). Bentham further attached numerical values to each of the seven criteria. These numerical values used a scale that for our purposes could range from –100 to +100. The negative values indicate levels of pain, whereas the positive values indicate levels of pleasure. Thus –100 represents extreme pain, as opposed to +100 that would indicate extreme pleasure. Once the scale is determined, one can perform a pain–pleasure calculation for alternative courses of action. Bentham suggested that we informally all undertake this calculation when we undertake moral decision-making and use it to decide upon a course of action. In common parlance it could be said that we undertake the equivalent of an assessment of pros and cons of possible actions and their outcomes when presented with a moral dilemma/conflict.

Table 4.2: Bentham's hedonic calculus.

CRITERIA	WHAT WILL BE MEASURED?
INTENSITY	How strong will the pleasure or emotional satisfaction be?
DURATION	How long will the pleasure last? Will it be short-lived or long-lasting?
CERTAINTY	How likely is it that pleasure will be the result/consequence of the course of action?
PROPINQUITY	When will the pleasure occur? In the short term or the long term?
FECUNDITY	How likely is it the action will produce more pleasure/pain in the future?
PURITY	How likely is the course of action to be accompanied by some pain? Is there some 'bad' that you need to accept alongside the 'good'?
EXTENT	How many people will be affected by the course of action to be taken?

However, the hedonic calculus, whilst providing a method of measurement that can be applied to real-life examples, for example work from home vs work in the office, it unfortunately proved problematic due to its subjectivity, because one is dealing in relative subject matter when assessing pleasure and pain, and therefore the measurement stands accused of lacking objectivity and empiricism. An argument has been put forward that Bentham's hedonic calculus is a forerunner of modern game theory. However, in asserting this, what is often forgotten in making this comparison is that game theory is a branch of probability theory that is irrelevant to moral philosophy as it cannot inform our sense and understanding of pleasure, happiness, pain or suffering.

ACT UTILITARIANISM AND RULE UTILITARIANISM

Act Utilitarianism and Rule Utilitarianism were developed in an effort to overcome the measurability issue that beset this ethical approach. It was also developed to counter the criticism brought against Utilitarianism that it justifies any action as long as it has the *'greatest good'* in terms of consequences (outcomes) rather than other actions that have been considered. In support of this position, it has been argued that unethical acts such as cheating, lying, stealing and breaking promises could be justified within the theoretical framework of Utilitarianism on the grounds they maximise happiness in a particular case. In order to try and correct this identified weakness in the theoretical framework, Bentham hypothesised the following about Act Utilitarianism and Rule Utilitarianism.

Act Utilitarianism has been explained in depth by Mel Thompson (2010) and is possibly best understood as the judging of the moral worth (value) of an action according to how well it generates the greatest good for the greatest number. In practice this equates to considering the consequences of each act separately.

The alternative to this approach would be:

Rule Utilitarianism (which is a limited position), also explained by Mel Thompson (2020), and best understood as the judging of the moral worth of an action to see how well it conforms to the moral rules that have been accepted according to the Utilitarian stance. In other words, this approach proposes that consideration should be given to the consequences (outcomes) of an act, if it were to be performed as a general practice without exception.

The difference between these two approaches is arguably summed up if one were to ask the question *'What if everyone did that?'*

STRENGTHS OF THE UTILITARIAN APPROACH

- The Utilitarian approach aims to maximise the utility, pleasure, happiness or common good of all.

- The Utilitarian approach extends to all living things (this approach is used within environmental ethics). The criterion that Bentham was said to have used in this regard was 'Can they suffer?'

- Utilitarianism concentrates on the scope ('bigger picture') rather than the microscope ('smaller picture'). This has particular benefits within the area of political and social reform.

- The Utilitarian approach has been seen to be compatible with the financial tool of cost–benefit analysis, which has proved to be the means of justification for certain actions within a capitalist context. This is especially evident when assessing the 'cost(s)' of a course of action versus the benefits of the expected outcome.

- Utilitarianism has been described as a practical ethical approach. It is not just a theoretical framework as it can be quite easily applied.

WEAKNESSES OF THE UTILITARIAN APPROACH

- Utilitarianism does not allow for situations/contexts in which it is difficult for us to know the exact consequences of an action.

- The utility of an individual can be marginalised and at worst disregarded, due to the Utilitarian premise of providing 'the greatest good for the greatest number'. In other words, it is a majority-based approach.

- The measurement of utility remains an obstacle for Utilitarianism. Not least because we have to ask ourselves **who** decides **what** measurement is to be used and **how** is the assessment to be made.

- The universality and egalitarian stance of Utilitarianism with regard to pleasure and pain has been questioned. Do we, or can we, **all** feel the same pain? Are pain levels consistent or are they a function of both our **biology** and **biography**?

UTILITARIANISM AND ENGINEERING

Interestingly the Utilitarian approach is often used by engineers when undertaking projects in a community context. The engineering decision is often made with the ***greatest good for the greatest number'*** being a moral maxim and on close inspection can be seen as an ethical motivator in engineering projects such as road building, dams, power generation and the like. Cost–benefit analysis is also a tool used, not just by engineers but within commerce and industry, as a measure of the principle of utility in which costs are seen as a measure of pain. The greater the costs, the less utility (happiness) is derived from the action. However, a cost–benefit analysis is a notoriously inaccurate measure of utility as it can often disregard feelings and does not provide for the fact that not all costs are necessarily known and therefore cannot be calculated. This is apparent when issues such as environmental damage, sustainability and substituting human tasks with technology (robots) are considered. The 'cost' of feelings, especially those associated with human dignity, freedom and lifestyle, are almost impossible to calculate within the context of engineering activity and we are once again left asking questions concerning the accuracy of a universal measurement in terms of utility/happiness/pleasure.

References

Mill, JS & Bentham, J (2004) *Utilitarianism and Other Essays*. Edited by A Ryan. UK: Penguin Classics.

Mizzoni, J (2010) *Ethics: The Basics*. UK: Wiley-Blackwell.

Thompson, M (2010) *Understand Ethics. (Teach Yourself)*. UK: McGraw-Hill Companies Inc.

LEARNING OUTCOMES

- In what ways does the philosophical framework for Utilitarianism differ from the classic ethics tradition?

- Describe the difference between Act Utilitarianism and Rule Utilitarianism.

- Explain how dualism and universalism are important aspects of the Utilitarian tradition.

- Outline what is meant by the 'common good' within the Utilitarian framework.

- Does Bentham's device of the 'hedonic calculus' advance the measurability of happiness and pain?

CHAPTER

5

HUMAN DIGNITY AND THE RIGHTS TRADITION

The scope and intent of the ethical approach of human rights cannot be fully appreciated without a comprehensive understanding of human dignity. The latter is axiomatic to both the understanding and the application of the human rights approach, which has gathered both traction and momentum in contemporary societies around the globe over the past 75 years.

WHAT FAMOUS PEOPLE HAVE SAID ABOUT HUMAN DIGNITY

> So what we're talking about here is human rights. The right to live like a human. The right to live, period. And what we are facing in Africa, is an unprecedented threat to human dignity and equality.
>
> – Bono

> When we achieve human dignity for all people – they will build a peaceful, sustainable and just world.
>
> – Antonio Guterres

> When it comes to human dignity, we cannot make compromises.
>
> – Angela Merkel

> I've always tried to avoid politics because most politicians that I know are quite dirty in terms of human dignity, ethics and morals.
>
> – Steven Seagal

It can be accepted that there is a price for a man's work, time spent or service done, but never for human dignity.

– Eraldo Banovac

The basic tenet of black consciousness is that the black man [*sic*] must reject all values systems that seek to make him a foreigner in the country of his birth and reduce his basic human dignity.

– Steve Biko

Human rights rest on human dignity. The dignity of man [*sic*] is an ideal worth fighting for and dying for.

– Robert Maynard

(BrainyQuote®, 2001–2022)

As can be seen, implicit in these quotes from famous people is an understanding of the interdependent relationship between human dignity and human rights in which the one, human dignity, informs the other, human rights.

HUMAN DIGNITY

These two words, 'human' and 'dignity', whilst having the ability to stand alone, can in combination present a powerful concept. Whilst the understanding of the word 'dignity' can be considered relative, within the context of human rights it has a universal meaning in contemporary times.

The word 'dignity' has its roots in the Latin word *'dignitas'*, which can be interpreted as status, excellence and/or respect. When translated from the Greek, the word 'dignity' suggests a form of worth or value and a guarantee of status in and through authority given. This was the position adopted by Aristotle with regard to human dignity, which he proposed was acquired through virtue and as a result conferred due to the hierarchical social status in ancient Greece in which statesmen were considered to acquire dignity and were accorded 'status' and 'worth' within the *'demos'*. It should therefore come as no surprise that against this social backdrop there was no notion of equal human dignity and women, children and slaves were denied it because of their position within the social hierarchy of ancient Greece.

However, the Greek translation of the word 'dignity' also suggests there is an incomparable value and entitlement to respect for human beings that arguably adopts a unique significance and the beginnings of what we would now interpret as **anthropocentricity**. Moreover the Greek translation of the word 'dignity' also speaks to an air of mysticism or spiritualism all its own, as both Greek citizens and Greek philosophers were beginning to realise.

MEDIEVAL DEFINITION AND APPROACH TO HUMAN DIGNITY

In Medieval Western philosophy we witness the immense significance and importance of theology, and in particular the Christian faith, specifically within the discipline of moral philosophy. Never is this more apparent than when examining the concept and reality of human dignity. The Italian theologian/philosopher Thomas Aquinas (1225–1274), who was a Dominican friar and priest, was the author of many works spanning law, politics, morality, justice and prayer. However, arguably his two best-known scholarly works are *Summa Theologica* and *On Essence and Being*, in which he provides arguments for the synthesis of Aristotelian philosophy and Christian theology and offers arguments for the existence of God that influenced Roman Catholic doctrine and tradition for centuries. In these works, Aquinas confirms the Christian concept of human dignity in the Medieval tradition that was intertwined with theology and expressed it through the framework he developed for Natural law. In this approach Aquinas defined humankind as having a rational nature that was conferred by God and which presented as free will and personal autonomy in human beings. This understanding of humankind was further expanded upon by Aquinas to include the theological concept of *'imago dei'*, which stems from a biblical proposition in the Old Testament Book of Genesis 1: 26–27 (1952), which states that God said:

Let us make man in our own image, after our likeness . . . So God created man in his own image, in the image of God he created him; male and female he created them.

This Christian theological definition of a human being, when expanded into the reality and concept of human dignity, implies sanctity and the suggestion that dignity is intrinsic to humans and is perceived to be a *'gift'* from God. By extension and implication, the concept of *'imago dei'* leads us to the understanding that it is one of the essential pillars of Natural Law in that human beings have a unique value or worth conferred on them that is suggestive of divine grace and as such they are paramount amongst all other living things over which they have dominion. However, of importance is this Medieval definition of human dignity still subscribes to our existing definition of moral agency, as it supports the notion that human beings have free will and autonomy due to their ability to reason, that in turn provides them with the unique capacity to make their own ethical and moral decisions and act accordingly.

KANTIAN APPROACH TO HUMAN DIGNITY

Whilst Kant's theory of deontology was developed in the absence of religion, it nevertheless supports the Medieval theological notion of personal autonomy as being essential to moral agency and expands upon this basis to incorporate the notion of rationality (arguably another essential pillar in Natural Law) that Kant suggests is a universal attribute of human beings that uniquely permits them to undertake the 'imperatives' of duty. Kantian theory focuses on the intrinsic value of human dignity possessed by every human due their ability to reason. This moral position is shared and supported by Dworkin (1994: 236), who concurs that our understanding of human dignity and its contemporary meaning is essential to;

the intrinsic importance of human life [and requires that] people never be treated in a way that denies the distinct importance of their own lives.

In other words, both Kant and Dworkin recognise that human dignity is inviolable by way of our common humanity.

MODERN DEFINITION AND APPROACH TO HUMAN DIGNITY

Modern definitions of human dignity have moved away from the theological towards the secular in an effort to create a universal definition for human dignity that can also serve as an objective measure so that one can identify both its existence and absence in situations of moral dilemma and/or moral conflict. This modern definition and understanding of human dignity also led to its legalisation and institutionalisation, which has been the basis for the 1948 Universal Declaration of Human Rights (UDHR).

On this basis of a secular and universal definition of human dignity, the philosopher Mette Lebech (2004) proposes that human dignity;

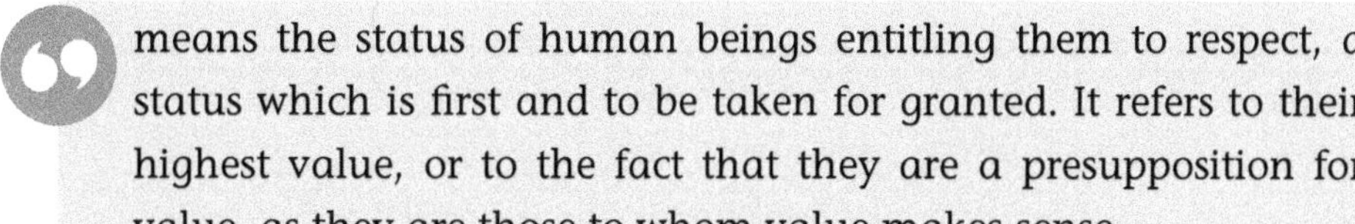

means the status of human beings entitling them to respect, a status which is first and to be taken for granted. It refers to their highest value, or to the fact that they are a presupposition for value, as they are those to whom value makes sense.

Within this definition we can observe that Lebech (2004) is reaching back into the Latin and Greek origins of the word 'dignity' so that she can create a universal and secular definition of this most difficult of concepts that remains devoid of any religious or spiritual frameworks. In this vein, she continues to clarify that:

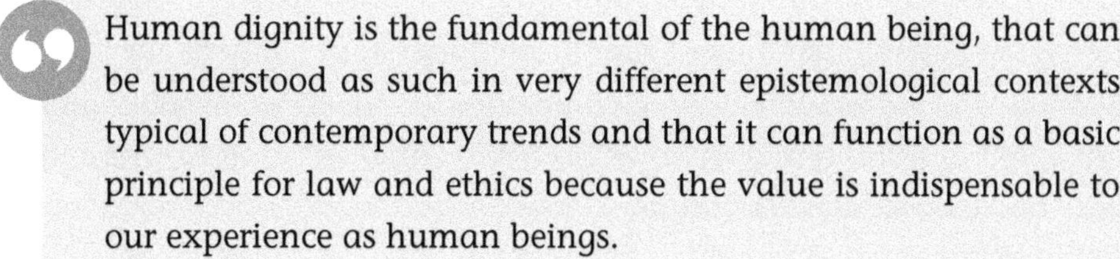

Human dignity is the fundamental of the human being, that can be understood as such in very different epistemological contexts typical of contemporary trends and that it can function as a basic principle for law and ethics because the value is indispensable to our experience as human beings.

However, it is arguably the modern philosopher Andorno (2007) who provides us with one of the finest contemporary meanings for human dignity when he states:

> Human dignity captures the notion that every human being is uniquely valuable and therefore ought to be accorded the highest respect and care.

It is perhaps by understanding this vision for human dignity and the universality of this definition, which has equality and egalitarianism implicit in its meaning, that we can fully and unquestionably understand both our worth and status as human beings and how this translates into an intrinsic notion of dignity for ourselves and others.

However, the scientific discoveries and 'new' technologies of the present have directly confronted Adorno's definition of human dignity and our understanding of it, to the extent the American scholar Robert Kraynak (2003: 9), a distinguished theologian, philosopher and political scientist, has unequivocally stated that in his view the challenge of our times is that we must recover a 'true and authentic understanding of human dignity', which is threatened by science and the technologies used in modern civilisation. He suggests;

> the dignity of man (sic) is not "a thing of the past", as many fear, but a living issue of primary importance to all of mankind as well as to citizens of modern democratic society.

CHALLENGES TO HUMAN DIGNITY

Contemporary times bring significant challenges to our understanding of human dignity. It is argued this has been a consequence of the rapid rate of scientific discoveries and technologies, that when used and applied, create previously unencountered moral dilemmas and conflicts for both individuals and wider society. Such challenges to our understanding and the meaning of human dignity have been seen to arise when examining subjects and case studies relating to:

- Abortion

- Immigration/Migration

- Cloning

- Euthanasia

- Foetal/stem cell tissue research

- Surrogacy

- Discriminatory practices

In each of these moral dilemmas we encounter challenges to both human dignity and human rights to the extent that Witte (2003: 121) says that some moral dilemmas have created moral conflicts in their wake to the extent that;

the corpus of human rights has become swollen to the point of eruption – with many recent rights claims no longer anchored in universal norms of human dignity, but aired as special aspirations of an individual or a group.

THE 'RIGHTS' APPROACH

The tradition of 'rights' has adapted through history. Examples of rights traditions are Athenian citizenship, the feudal system, the class system and political revolution. The existence of these historical examples and the 'rights' accorded in general within them are accepted to be the product of moral, social and political agreements between human beings. *The Oxford Dictionary of Philosophy* (2008: 318) describes them as follows:

Rights are frequently held to 'trump' other practical considerations, which requires seeing them as not themselves simply grounded in the interests of the right-holder, but perhaps existing in virtue of more central considerations of the duties we owe to each other.

This definition confirms that, at their simplest, 'rights' can be legal or moral and are best interpreted as *'entitlements'* aimed at universality owing to the fact they apply to everyone in that situation. Rights are essentially concerned with the idea of an ideal world in which the values and worth extended to human life are unequivocal and universal. At its simplest, if we were to scrutinise the rights tradition and approach, we should be examining the conditions of form and structure regarding those rights (see Figure 5.1).

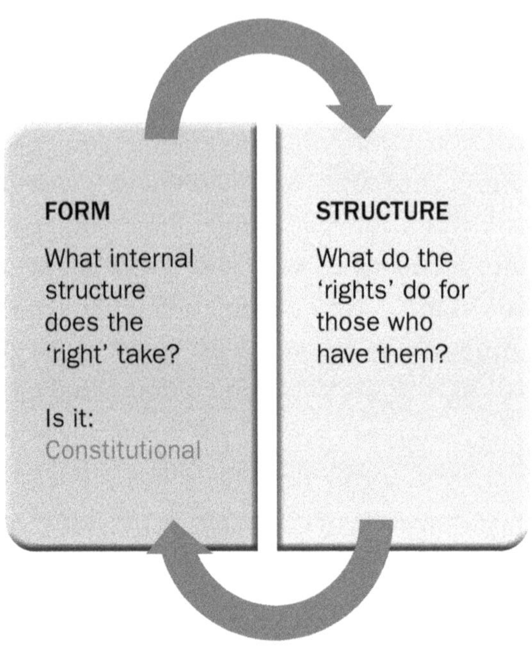

Figure 5.1: The rights tradition.
Source: Author's own.

HISTORICAL EXAMPLES OF RIGHTS

As we have learned, 'rights' can be legal and/or moral and aim at universality. We have come to associate them with constitutions and legal documents that aim to 'guarantee' freedoms. One of the earliest examples of this is considered to be Britain's Magna Carta of 1215, drafted by the Archbishop of Canterbury as a Royal Charter of 'rights' (also known as the Charter of Liberties), guaranteeing specific 'rights' and equality for all before the law. It was signed by King John at Runneymede, is recognised as a founding document for a national constitution and has been used subsequently as a template for national constitutions in which the basic 'rights' of citizens are both outlined and guaranteed by law. Most notable examples of legal documents embodying the 'spirit' of the law first witnessed in Magna Carter are the American Constitution and our own South African Constitution, developed for our new democracy and in which the 'rights' of all citizens are clearly stated. Therefore, the respective governments of all and any nation with a constitution have the duty of upholding the rights afforded its citizens by law.

The 'rights' tradition has occupied a very special place in philosophical thought, not least because it can be seen to overarch not just philosophy, but moral philosophy, politics, law and economics. John Locke (1632–1704), the philosopher (commonly known as the father of liberalism), is well known for liberal thought and his position that human beings have three natural 'rights', namely *life*, *liberty* and *property*. However, it is perhaps the Utilitarianist John Stuart Mill (1806–1873) whose views in his book *On Liberty* (1859) that resonate most accurately with our current understanding of the rights tradition. He says:

> When we call anything a person's right, we mean that he [*sic*] has a valid claim on society to protect him [*sic*] in the possession of it, either by the force of law, or by that of education and opinion . . . To have a right, then, is, I conceive, to have something which society ought to defend in the possession of.

Millsian philosophy speaks directly to governments as having the responsibility of both ensuring and upholding, and ultimately defending, the human rights of their citizens which have been constitutionally guaranteed.

THE 1948 UNIVERSAL DECLARATION OF HUMAN RIGHTS (UDHR)

Our modern approach and understanding of human rights originated as a response to the atrocities witnessed in the Second World War. Once uncovered, the extent of the war crimes that were committed (some of which were brought to light during the Nuremberg Trials) created a groundswell of opinion that suggested a legal agreement be established in the form of a human rights initiative that would result in a legal document, binding upon signatories, that would ensure the rights, dignity and freedom of the individual would always be protected and defended.

Eleanor Roosevelt (1884–1962), the wife of the former American President Theodore Roosevelt, was a liberal, who actively campaigned for social reform and a global recognition of the need for a universal charter on human rights to ensure the atrocities committed against individuals in the Second World War would not be able to happen again.

Speaking at the Conference for Human Rights in La Sorbonne in Paris (1948) she said:

> People who have glimpsed freedom will never be content until they have secured it for themselves . . . People who continue to be denied the respect to which they are entitled as human beings will not acquiesce forever in such denial.

We can see in her words the interrelationship – even interdependency – and inviolability of the concepts of human dignity and human rights which she championed in her post as the Chairperson of the United Nations Human Rights Committee. She eloquently shared the vision and rationale of a Human Rights Charter when she stated:

> The field of human rights is not one in which compromises on fundamental principles are possible.

In adopting this position with regard to the importance of human rights, Eleanor Roosevelt was unequivocal in both the need for and the defence of basic human rights. Her famous speech sought not to limit the purview of human rights, evident in the connection she made between human dignity, free will, justice, peace and freedom. The latter, she insisted, was a basic human right and one that should not fall victim to social progress or *'better standards of life'* but should be promoted and encouraged for all.

In 1948 the first Universal Declaration of Human Rights (UDHR) was signed and its approved remit was that it should be the foundation of freedom, justice and peace in the world in keeping with Eleanor Roosevelt's position that fundamental human rights should be extended to every human being, all of whom she saw as being citizens of the world. Perhaps most important in achieving her goal of a Human Rights Charter was that at the core of this 'new' document, called the UDHR, was the universal notion of human dignity that it described in Article 1 as:

 All human beings are born free and equal in dignity and rights. They are endowed with reason and conscience and should act towards one another in a spirit of brotherhood.

The five basic human rights outlined in the 1948 UDHR can be seen in Table 5.1.

Table 5.1: The five articles in the 1948 UDHR relating to basic human rights.

ARTICLE	BASIC RIGHT CONFERRED
1	Right to equality
2	Right to freedom from discrimination
3	Right to life, liberty and personal security
4	Right to freedom from slavery
5	Right from freedom of torture and degrading treatment

When read and understood, the 1948 UDHR and the major reviewed version in 1966 and subsequent versions, can be seen to recognise the inherent dignity and equal and inalienable rights of all members of the human family. It can be seen to recognise not just civil and political rights, but those of a social and cultural nature too. Subsequent revisions of the UDHR have endeavoured to expand the scope of the rights tradition to include those of the environment, animals, climate, privacy and even space and the cosmos. Essentially it recognises our relationships one with another, with nature and with science and technology.

RIGHTS DUALISM

The rights approach, whilst rooted in a universal notion of dignity, has the dualism of duty in its philosophical framework. One cannot be given a 'right' without a corresponding duty to uphold it and defend it. An example of this would be that whilst we might have the 'right' to free speech, we also have the responsibility not to take part in hate speech;

we have the 'right' to freedom from torture, but the responsibility not to commit it against others.

America is currently experiencing the reality of this dualism between rights and duties in its constitution, specifically with regard to the First Amendment of the Declaration of Independence in which citizens have the right to bear arms, but the responsibility of securing weapons when not in use and of being of 'sound mind' should they ever commission them. The number of arbitrary mass shootings in contemporary America is causing increased concern amongst both politicians and citizens as they witness (unfortunately in many cases first-hand) the abuse of this fundamental constitutional right to bear arms. There are calls by anti-gun activists and lobbyists that the First Amendment right to bear arms be rescinded to prevent further unnecessary loss of innocent lives and the lack of dignity that corresponds with this for the victims and families of such mass shootings.

HUMAN DIGNITY AND HUMAN RIGHTS IN ENGINEERING

Human dignity and human rights are first and foremost concerns and considerations for engineers who work in a social context. South Africa re-joined the United Nations post-1994 and since then has made a significant contribution in the arena of human rights in addition to being instrumental in developing a refined sense of human dignity for the African continent. Not only has the South African Constitution been hailed as one of the most progressive in the world, but in the South African Human Rights Commission (SAHRC), established in 1995, South Africa has created and developed an independent and progressive body which serves to strengthen South Africa's constitutional democracy. The role and function of this organisation is to;

support the constitutional democracy through promoting, protecting and monitoring the attainment of everyone's human rights on South Africa without fear, favour or prejudice.

Additionally, this august body also has the remit to;

 take steps and secure appropriate redress where human rights have been violated.

For engineers working in a social context within South Africa, human rights and human dignity should be uppermost when considering and/ or undertaking engineering projects and engineering activities, not least because if they are disregarded, legal action can be taken through the channels of the SAHRC if violations are considered to have taken place.

Within South Africa's borders there is the additional consideration that must be given to the legal jurisdiction extended to Tribal Councils in rural areas. These are often the source and support of the cultural and traditional notion of both human dignity and human rights and they are a recognised grouping within South Africa's legal system.

Finally, as engineers working in a social context within South Africa, there are rights that must be upheld in the profession and the engineering industry. These extend to the right to health and safety in the engineering workplace for employees and for the public who are being served by the engineering activity. Engineers must also ensure the integrity of an engineering project and/or activity so that it does not conflict with or deny existing human rights. This most often needs to be considered when engineering projects and/or engineering activities have environmental impact, especially when considering activities such as mining, fracking, the use of scarce natural resources such as water, heritage sites and the encroachment on wildlife conservancies, or areas of natural beauty.

WEAKNESSES OF THE RIGHTS TRADITION

- Legislation does not always ensure rights.
- Rights do not always coincide with justice.
- Relativity is applied to rights, for example shelter, education, torture.
- Rights are geared to minimum standards not maximum standards.
- Is equality achieved or practical in the rights approach?

STRENGTHS OF THE RIGHTS TRADITION

- The universality of rights.

- Rights can be actively challenged and changed.

- Rights can provide a legal guarantee of/for moral status.

- The rights tradition is an active form of applied ethics and can be used together with other ethical traditions, specifically deontology.

- The UDHR can cause non-signatories who breech the charter to be challenged, for example Soviet Union and Saudi Arabia.

References

Andorno, R (2007) Global bioethics at UNESCO: in defence of the Universal Declaration on Bioethics and Human Rights. *Journal of Medical Ethics* 33: 150–4. Also available at **https://www.researchgate.net/publication/24350326_Human_Dignity_and_Human_Rights_as_a_Common_Ground_for_a_Global_Bioethics**

Bible, The. Revised Standard Version (1952) UK: WM Collins, Sons & Co Ltd.

Blackburn, S (2008) *The Oxford Dictionary of Philosophy*. UK: Oxford University Press.

BrainyQuote® (2001–2022) Famous People Quotes on human dignity available at **https://www.brainyquote.com/topics/human-dignity-quotes#:~:text=Freedom%20is%20the%20open%20window,human%20spirit%20and%20human%20dignity. &text=When%20we%20achieve%20human%20rights,%2C%20sustainable%2C%20and%20just%20world. &text=When%20it%20comes%20to%20human%20dignity%2C%20we%20cannot%20make%20compromises**

Dworkin, R (1994) *Life's Dominion. An Argument About Abortion, Euthanasia and Individual Freedom.* New York: Vintage, p 236.

Kraynak, R (2003) *Defending Human Dignity: The Challenge of our Times in Defense of Human Dignity.* Essays for our times. Edited by Robert Kraynak & Glen Tinder. U.S. University of Notre Dame Press.

Lebech, M (2004) **https://mural.maynoothuniversity.ie/392/1/Human_ Dignity.pdf**

Mill, JS (1859) *On Liberty*. Amazon (Kindle Version 2002).

Roosevelt, E (nd) **https://www.americanrhetoric.com/speeches/eleanor- roosvelt.htm**

South African Human Rights Commission. **https://nationalgovernment. co.za/units/view/61/south-african-human-rights-commission-sahrc**

Universal Declaration of Human Rights available at **https://www.ohchr.org/ sites/default/files/UDHR/Documents/UDHR_Translations/eng.pdf**

Witte, J (2003) *Between Sanctity and Depravity: Human Dignity in Protestant Perspective in Defense of Human Dignity*. Essays for our times. Edited by Robert Kraynak & Glenn Tinder. US: University of Notre Dame Press.

LEARNING OUTCOMES

- Explain the need for a Universal Declaration of Human Rights in 1948.

- Outline the dualism that exists with the rights tradition.

- Discuss whether we can realistically legislate for an ideal world.

- Rights do not always coincide with ethics, for example the 'right to life', but in the case of some signatory countries the law can provide for a death penalty. Discuss.

- What right would you add to the current UDHR and explain why.

THE AFRICAN ETHICS APPROACH

6

There has been a realisation, some would say awakening, to the idea that philosophy extends beyond its original Western roots, contexts, paradigms and constraints. This is especially true of moral philosophy (ethics) where we see the discipline embracing Eastern, Middle-Eastern, Chinese, Indian, indigenous and African ethical traditions and approaches as the authenticity and practicality of ethics becomes central to the flourishing and well-being of all around the globe. The contemporary study of non-Western philosophies also serves to make a strong case for ethical relativism, as will be seen later in this chapter.

Recognition and academic study of these non-Western ethical traditions has served to differentiate the previous 'global' universal ethic of the 'West' from those in societies that lie beyond this geopolitical boundary. It examines and clarifies the scope of differing values and beliefs (ethics). Moreover, since the professional engineering workforce has become increasingly international, exposure to these non-Western ethical traditions and approaches and an understanding of them has become important for engineers within their professional capacity, as their work is undertaken within a social context.

What has become apparent from the study of non-Western moral philosophies and approaches is that, even though there are significant differences in their constructs, there remains throughout a common notion of human dignity where often the spiritual and/or mystical becomes important to the understanding of not just human dignity and personhood but also to the role of human dignity within the workplace and the impact this can exert on engineering projects/activities that might be undertaken therein.

NON-WESTERN ETHICAL TRADITIONS

In trying to study both the differences and similarities between Western and non-Western ethical traditions and approaches, the following terms are used and their impact, scope and extent examined as a means of common measure by scholars:

- Epistemology

- Indigenous Knowledge (IK)

- Ethnophilosophy

EPISTEMOLOGY

The Oxford Dictionary of Philosophy (2008: 118) defines epistemology as:

The theory of knowledge. Its central questions include the origin of knowledge; the place of experience in generating knowledge and the place of reason in doing so . . . All these issues link with other central concerns of philosophy, such as the nature of truth and the nature of experience and meaning.

Broadly speaking the narrow definition of epistemology is arguably the study of knowledge and justified beliefs. Traditionally Greek philosophy, especially when considering Platonic philosophy, proposes that knowledge comprises true belief and logic, but this view is perhaps too simplistic, given the role that beliefs occupy in the non-Western ethical tradition. The significance of beliefs in non-Western ethical traditions and approaches, especially when undertaking an examination of African ethics, can be viewed as central to moral reasoning and decision-making on the African continent.

The broader definition of epistemology, provided in *The Oxford Dictionary of Philosophy*, unlike its 'narrow' Classical counterpart, extends to include the creation and dissemination of recorded knowledge that can be in the form of either oral or written tradition and it is here where we see the relationship with indigenous knowledge.

INDIGENOUS KNOWLEDGE

Indigenous knowledge (IK) has been defined by Unesco (nd) as follows:

Local and indigenous knowledge refers to the understandings, skills and philosophies developed by societies with long histories of interaction with their natural surroundings. For rural and indigenous peoples, local knowledge informs decision-making about fundamental aspects of day-to-day life. This knowledge is integral to a cultural complex that also encompasses language, systems of classification, resource use practices, social interactions, ritual and spirituality.

The World Bank (2007) has undertaken significant research in the area of IK, however it is Warren's (1991) definition of IK that, whilst it can be seen to concur with that of Unesco, has been expanded and compared with traditional systems of epistemology in Western societies. Warren defines IK as follows:

Indigenous knowledge (IK) is the local knowledge – knowledge that is unique to a given culture or society. IK contrasts with the international knowledge system generated by universities, research institutions and private firms.

Using these two definitions to provide meaning to the term 'IK', African ethics can be seen to conform in large part to these two meanings in that it relies on its own particular system of IK.

ETHNOPHILOSOPHY

The term 'ethnophilosophy' has been defined by the *Routledge Encyclopedia of Philosophy* (Karp & Masolo, 1998) as;

bodies of belief and knowledge that have philosophical relevance and which can be redescribed in terms drawn from academic philosophy, but which have not been consciously formulated as philosophy by philosophers. These bodies of belief and knowledge are manifested in the thoughts and actions of people who share a common culture.

This definition would appear to support the notion of indigenous philosophical systems and is used most often within the context of African philosophies and African ethics.

WHAT IS AFRICAN ETHICS?

Janz (2002), in writing about Africa and its unique philosophy and ethics frameworks, coined the phrase 'philosophy in place' and provocatively suggests that:

> African philosophy is often thought of as a location – a space to be carved out on the map of global philosophy, a territory distinct and apart from Western thought.

However, whilst endorsing the relationship between philosophy (and moral philosophy) and a distinct geographical location, he does not present a clear understanding of an African philosophical framework and/or its applications in the way that certain African philosophers have been able to do, who share their African language, culture and beliefs in common.

A distinguished scholar in the field of African ethics, Molefe (2016) shares his understanding of African ethics not just in terms of its geography, but also hints at its possible applications when he says:

> I use the notion African ethics to refer to general and salient moral intuitions that are considered to be salient below the Sahara.

Gyekye (2010: 1) arguably provides us with a definition of African ethics that is significantly more encyclopaedic when he proposes that:

> African ethics is used to refer to both the moral beliefs and presuppositions of the sub-Saharan African people and the philosophical clarification and interpretation of those beliefs and presuppositions.

From both definitions we are left with the distinct impression that African ethics has a philosophical framework that is unique, due in no small part to its geographic region. What will also become apparent is that, according to one of the pre-eminent scholars in the study of African Ethics, Munyaradzi Felix Murove (2009: 26)

African ethics is one of the world's ethical traditions with its own contribution to make towards a global ethic.

THE PARTICULARITY OF AFRICAN ETHICS

When examined, African ethics (sub-Saharan Africa) can be seen to consist of the following four elements that are central to its philosophical framework and provide it with an explication of its distinct position in terms of its moral philosophy:

1. **Integral spirituality**

   The element of integral spirituality in African ethics results in an 'holistic worldview'.

2. **Inclusive community**

   The element of inclusive community includes the living dead (ancestors), the living, the yet to be born, and all of creation.

3. **Interdependence and connectedness**

   African ethics tradition applies to all living things. It recognises the 'web of life' and the way in which an interconnectedness exists based on the interdependence of beliefs and evidence of both the seen and unseen.

4. **Communal ethics**

   African ethics tradition has a very real understanding and pragmatism with regard to the *'common good'*. As such, there is a sense of egalitarianism and utilitarianism that threads its way through its distinctive accepted practices and norms (morals) and beliefs (ethics).

However, whilst these four elements are key to an understanding of African ethics, it is by examining their application that we can begin to understand the way in which they shape both the ethics and morals of indigenous South Africans and provide them with a very distinct ethical tradition.

AFRICAN ETHICS UNPACKED

African philosophy and language were traditionally oral. This has led to the criticism that because historically neither philosophy nor language was recorded in writing, both could be accused of being subject to memory. Thus African ethics has been accused of being a *'morality of memoria'* in the absence of a written epistemology. A written language was developed that has subsequently needed interpretation for understanding and meaning to flow. Both the written language and the interpretation of African ethics has been spearheaded by the West/colonials and this act brings into question the authenticity of the philosophical frameworks associated with this regional/continental ethic and its interpretation. The question that pervades any discussion surrounding African ethics continues to be: has the philosophy and ethical tradition(s) attributed to African ethics been interpreted through the lens of Western frameworks and has this been instrumental in driving the persistent question directed towards it? This initial question has surely been the touchpaper for the question that almost takes on the substance of a statement, which is: *'Is African ethics a philosophy or a practice?'* The validity of this question has been cemented in large part by the notion and stance of Ubuntu, which, whilst forming the beating heart of African philosophy and African ethics, is a concept that can transcend geographical borders and many would say provides the basis of a *'lived'* philosophy.

UBUNTU

This unique stance of African ethics has been recognised globally and underscored the 'new' democracy in South Africa. It has proven difficult to define and indeed some would argue that it needs to be experienced to be fully understood and for it to have and share a common meaning.

Within its African context, Ubuntu is *'Umuntu ngumuntu ngabantu'*. This phrase translates as *'A person is a person through persons'*. Considering this translation, African philosophers can agree on a common, some might say universal, interpretation of Ubuntu, which is *'I am because we are'*.

It has been suggested this universal interpretation of Ubuntu might also be understood by the phrase *'We all depend on one another for our existence'*, a position that was further clarified by the South African struggle hero Steve Biko, remembered in a speech given by a former President of South Africa Thabo Mbeki (Warby, 2007), who stressed that within South Africa;

the cornerstone of society is man [*sic*] himself.

One can see through these interpretations a consensus in recognising the three key aspects central to the understanding and application of African ethics: interconnectedness; interdependence; and inclusive community, suggesting a communal ethic alongside the underlying suggestion of an integral spirituality that accompanies all three.

FURTHER INSIGHTS INTO UBUNTU

Broodryk (nd) has sought to define the attributes and meanings of Ubuntu through the use of an acronym (see Table 6.1) which would appear not only to identify Ubuntu's key attributes and principles but would also seem to capture both the meaning and spirit that informs this uniquely African ethical tradition.

Table 6.1: The positive attributes and meanings of African Ubuntu philosophy.

UBUNTU ATTRIBUTE	AFRICAN UBUNTU MEANING
U = UNIVERSAL	Global, intercultural brotherhood
B = BEHAVIOUR	Human (humane), caring, sharing, respect, compassion (love, appreciation)
U = UNITED	Solidarity, community, bond, family
N = NEGOTIATION	Consensus, democracy

UBUNTU ATTRIBUTE	AFRICAN UBUNTU MEANING
T = TOLERANCE	Empathy (forgiveness, kindness)
U = UNDERSTANDING	Understanding

There are further distinct differences between African ethics and its Western philosophical counterparts. Perhaps the sharpest of these differences is that within African ethics the concept and reality of *'self'* is perceived and present outside of the body and open to all. This differs vastly from Western philosophies that teach us that *'self'* is something private and hidden, vests within our bodies and is not to be shared without cause. African ethics also teaches that a person only exists in relation to other persons and their surroundings and that these relationships contribute to the sum of the person. This can be interpreted as the *'web of life'* that is central to the African ethical tradition.

THE KEY ELEMENTS FOUND IN THE FRAMEWORK OF UBUNTU

Gyekye (2010) reminds us that in many African languages there is no word for *'ethics'* or *'morality'*. Whilst it may be true there is no direct translation for these words, there is nevertheless an understanding of them which is experienced through the human character, much as Aristotelian virtue theories. Gyekye (2010:3) reminds us that in African ethics;

character is used to refer to what others call 'ethics' or 'morality'.

The importance of a person (or umuntu), whilst central to the tradition of African ethics, can be seen to extend and be reliant on other aspects of African life which together serve to enrich this ethical approach. This has been identified by Munyaka and Motlhabi (2009) and modified by the authors into four key elements that can arguably be interpreted as central to Ubuntu and the concept of the person, which together can be seen as providing the all-important building blocks in the philosophical framework and tradition of African ethics.

Individualism

Ubuntu has a person at its core who is considered and treated as an 'end in themselves'.

Isidima

This translates as 'dignity' and acknowledges that within the tradition of Ubuntu, life is the greatest gift bestowed on humanity by the divine and as such is to be respected. This recognises that a person is not a thing nor a number, but has intrinsic value and is worthy of respect. There is a sense that human dignity provides a sense of the divine in a person.

Mutuality

The interdependence that flows through Ubuntu is a central pillar to both its demonstration and understanding. Mutuality comes from this, as does a sense of 'belonging' for 'persons'.

Relationships

Ubuntu has an implicit understanding of the importance of relationships within the life of the individual and the significance of this when it comes to the aspect of mutuality. The latter is achieved, not just in the sense of our relationships with other human beings and their relationship with ourselves, but also through our relationship(s) with the environment (both land and cosmos), animals and our spiritual /mystical creator, both directly and indirectly. Ubuntu proposes this to be a vital life force as it speaks to participation and the interconnectedness of life. To this end, Ubuntu supports the belief that a person has no independent existence; rather their existence is interwoven with the rest of the universe. To support this contention within its philosophical framework and tradition Ubuntu disregards the Western notion of *'ego'* and its accompanying *'individualism'* in favour of an egalitarian communalism formed through relationships.

STRENGTHS OF UBUNTU

- Ubuntu vests in a strong social consciousness. Its focus is living in relation to others (**relationality**) and directly involves the person in social and moral roles, duties, obligations and commitments.

- Ubuntu is life centred.

- Ubuntu is inclusive and has a profound sense of egalitarianism running through it.

- Ubuntu is community centric. Léopold S Sengor, the first president of the Republic of Senegal, used the term *'communalism'* for this type of ethics approach. Historically, Africa's rural living and accompanying tribal social structures encouraged this. Considering this, it is argued that Ubuntu is best suited for rural existence in which community is paramount and relationships are central.

- Tolerance, compassion and forgiveness are central to the ethical approach of Ubuntu.

- Nation-building fits the profile of Ubuntu with its emphasis on equality, egalitarianism and relationships.

- Ubuntu has no notion of egoism and does not subscribe to individualism, both of which in recent times have become synonymous with Western, capitalist and liberal democracies.

WEAKNESSES OF UBUNTU

- The philosophy of Ubuntu would appear to be at odds with both the concept and practice of capitalism. Rather the thrust of Ubuntu could be viewed as being that of socialism.

- African ethics and the theory of Ubuntu arguably lacks epistemology.

- Ubuntu has no notion of 'independent existence' and arguably abrogates a sense of 'personal responsibility' in favour of the collective.

- Both the theory and practice of Ubuntu do not consider urban lives in which a sense of community is marginalised and where neighbours often remain unknown to each other as urban life encourages independent existence at the expense of relationality.

AFRICAN ETHICS AND ENGINEERING

African ethics most frequently imposes itself within the social context of the practice of engineering in Africa. Proposed engineering activities/projects which become the subject of public engagement often must bridge the dual world of tribe and State initiatives and outcomes. Community engineering projects need to be mindful of the interconnectedness of life when locating engineering activities in rural areas in which both the land and certain tribal traditions are sacred.

The N2 toll road in South Africa has been subject to much community scrutiny, disapproval and stoppages due to the proposed route going over ancient, ancestral, tribal burial sites. Local tribes were not prepared to incur the wrath of their ancestors or sever ties with them because of the proposed toll road route and the disruption that would occur to the final resting places of family members, let alone the disruption to those buried there by the noise and impact of commercial vehicles 'driving over them'. Similar emotions were stirred when the Lesotho Highlands water project was undertaken and community opposition arose in light of the proposed water capture from local areas, which local communities believed would upset the time-honoured balance of life accepted and expected by their tribe. When such incidents occur it seems to highlight the controversy of whether African ethics is a philosophy or whether it is a 'practice' which is arguably supported by its narrative of 'return', since implicit in its framework is the acknowledgement of colonialism and that a 'return' to something and anything African is necessary in order for African society as a whole to prosper.

References

Blackburn, S (2008) *The Oxford Dictionary of Philosophy*. UK: Oxford University Press.

Broodryk, J (nd) **https://repository.up.ac.za/bitstream/handle/2263/28 706/04chapter4.pdf?sequence=5**

Gyekye, K (2010) African Ethics. *The Stanford Encyclopedia of Philosophy*. Edited by EN Zalta. Available at **https://plato.stanford.edu/entries/ african-ethics/**

Janz, BB (2002) The Territory is Not The Map: Place, Deleuze, Guattari and African Philosophy. *Philos Africana* 5(1): 1–18 Available at **https://www.jstor.org/stable/10.5325/philafri.5.1.0001**

Karp, I & Masolo, DA (1998) Ethnophilosophy, African. *Routledge Encyclopedia of Philosophy*. **https://www.rep.routledge.com/articles/ thematic/ethnophilosophy-african/v-1**

Molefe, M (2016) African Ethics and Partiality. *Phronimon* 17(2): 1–19. **https://dx.doi.org/10 17159/2413-3086/2016/142**

Munyaka, M & Motlhabi, M (2009) *Ubuntu and its Socio-moral Significance in African Ethics: An Anthology of Comparative and Applied Ethics*. Pietermaritzburg: University of KwaZulu-Natal Press, pp 63–84.

Murove, MF (2009) *African Ethics: An Anthology of Comparative and Applied Ethics*. Edited by Munyaradzi Felix Murove. Pietermaritzburg: University of KwaZulu-Natal Press.

Unesco (nd) **https://en.unesco.org/links**

Warby, V (2007) President Mbeki recalls Steve Biko's vision of Ubuntu. **https://www.skillsportal.co.za/content/president-mbeki-recalls-steve-bikos-vision-ubuntu**

Warren, DM (1991) *Using Indigenous Knowledge in Agricultural Development*. World Bank Discussion Paper No 127. Washington, DC: The World Bank.

World Bank (2007) **https://web.worldbank.org/archive/website00297C/ WEB/0__CO-23.HTM**

LEARNING OUTCOMES

- Explain why Ubuntu has been viewed by some as a rural, naive, philosophical approach not suited to modern, screen-based life.

- Was the TRC an outward manifestation of Ubuntu?

- Can Ubuntu be adapted to suit modern, urban life?

- Discuss why Ubuntu has been accused of being a Western construct developed from a European narrative.

- Is Ubuntu a philosophy or a practice?

ENVIRONMENTAL ETHICS

7

Whilst it is assumed by many to be a 'new' ethical approach, the history of environmental ethics stretches back to the Classical era in Western philosophy and has moved in tandem with the main philosophical approaches throughout the Medieval, Modern and Contemporary eras.

The Oxford Dictionary of Philosophy (2008: 116) tells us that:

> The central problem specific to thinking about the environment is the independent value to place on such things as preservation of species, or protection of the wilderness.

However, it is perhaps Sagoff (1993: 10) who provides us with the most accurate definition of our contemporary understanding of the word 'environment' and the sentiment we attach to it when he says:

> The environment is what Nature becomes when we view it as a life-support system and as a collection of materials.

We might all be familiar with buzz words such as:

- Greenhouse effect
- Acid rain
- Global warming
- Carbon footprint
- Conservation
- Recyclable

- Sustainability

- Bio-diversity

- Environmental justice

Their full meaning and our unhindered understanding of these terms have only been apparent in this millennium.

PHILOSOPHICAL APPROACHES TO THE ENVIRONMENT

The Classical era

Singer (1993: 268) aptly reminds us that:

> According to the dominant Western tradition, the natural world exists for the benefit of human beings.

Whilst this might be accurate, it is the difference in philosophical approaches over the past two thousand years that has brought us to modern times and our acute understanding of the need for protection and preservation of our natural world if it is to be sustained in the present and in the future.

Preservation and protection were not words used by Greek philosophers in the Classical era of Western philosophy when theorising about the relationship human beings had with their natural environment. Aristotle took an anthropocentric approach and assumed a natural hierarchy that attached superiority to human beings because of their ability to reason. This ascendancy provided human beings with an intrinsic authority over their natural environment that included all living things such as animals, plants, insects and fish in addition to non-living things such as minerals, oceans, rivers and the like. This belief in the ascendancy, authority and dominion of human beings over the natural world served to define relationships in the Classical era in Western philosophy. It not only shaped the relationship that humans had with each other (there was a rigid social hierarchy that conferred status by gender and social position) but also the relationship they had with the natural world. These dynamics continued into the Medieval era.

The Medieval era

The Medieval theologian/philosopher Thomas Aquinas is chiefly associated with developing the High Catholic Christian approach in which God made 'man' in his own image and likeness (*imago dei*) and the earth for 'His' purposes. Aquinas used the account of God's creation in Genesis 1: 28–30 as support when it tells us that God commanded 'man' to;

> be fruitful and multiply, and fill the earth and subdue it; and have dominion over the fish of the sea and over the birds of the air and over every living thing that moves upon the earth . . . [B]ehold, I have given you every plant yielding seed which is upon the face of all the earth and every tree with seed in its fruit; you shall have them for food.

Aquinas proposes the relationship with human beings, the natural world and God was both the product and the design of the divine. The interpretation of the first two chapters in the Old Testament book of Genesis once again supported the primacy of humankind and an implicit 'mastery' which ironically coincided with a duty of care for the natural world that was assumed and undertaken by human beings at the behest of God. Singer (1993: 268) provides us with an accurate sense of this Medieval relationship between humanity and nature when he states:

> Nature itself is of no intrinsic value and the destruction of plants and animals cannot be sinful, unless by this destruction we harm human beings.

This 'sanctioned' understanding of the human relationship with the natural world drove exploration and exploitation of resources and people during this Medieval era.

The Modern era

As discussed previously, the Modern era in Western philosophy coincided with the Age of Enlightenment in which science and discovery brought about a major change in both life and philosophy. The new *'rational approach'* developed by Kant, in which he credited human beings with rationality, the ability to reason and therein their moral autonomy, did nothing to dislodge an anthropocentric approach to human beings and their relationship with nature. Nature still *'belonged'* to human beings, who were uniquely placed to decide both its *'intrinsic'* and *'instrumental'* value. The latter became of increasing importance, as science provided human beings with the means to extract resources from the ground and use them in *'new'* manufacturing processes and technologies for their benefit. Therefore the natural world, whilst having a degree of intrinsic value, was fast becoming a resource of instrumental value and was the sole source of many of the resources it provided. Many have suggested the Modern era saw the birth of *'consumerism'* as the expansion of colonisation began to redefine geographic and political boundaries and the idea took root that anything in nature is all part of God's bounty to be used and bought by *'man'*.

Interestingly, there was increasing opposition to this view. Adam Smith in his seminal work *The Wealth of Nations* spoke about *'the tragedy of the commons'*, which referred to the continual exploitation of the world's natural resources. People were beginning to recognise that natural resources were finite and should be preserved or at least used in a sustainable manner. There was also an appropriate understanding of the *'force of nature'* during this era and how it could resist control by human beings and could also wreak havoc upon established societies. Never was this more evident than in the great earthquake and accompanying tsunami of Lisbon in the year 1755 which not only devastated the city, Portugal's economy and claimed more than a third of the entire population of this thriving Portuguese capital, but its shock waves were felt as far away as Morocco, where it has been estimated that up to 10 000 inhabitants were killed. This event caused scholars and influencers during this era to recognise the natural world would not always be passive and controllable and at the behest of *'man'*, which led to the question being posed of whether 'man' belonged to nature or nature belonged to *'man'*.

The Contemporary era

It is in this era that we begin to see a significant shift in our interpretation of both nature and our relationship with it that defied all earlier philosophical interpretations. Tolkien's (2013) famous novel *The Lord of The Rings* alludes to this shift in the relationship between human beings and the natural world evidenced in the words of Treebeard to Galadriel as they parted for the last time:

> The world is changed.
>
> I feel it in the water.
>
> I feel it in the earth.
>
> I smell it in the air.
>
> Much that once was is lost,
>
> For none now live who remember it.

This sentiment speaks to the way in which nature and the natural world is now celebrated for its intrinsic value and the dynamics of 'man's' relationship with the environment is no longer one of dominance, but rather one in which we recognise the integral part nature plays in our own survival. We are now intimately aware of the need for preservation, protection and sustainability as the finite resources of the natural world and their balance is unfolding and becoming better understood. Moreover, we are also now familiar with the potential destruction of both finite natural resources and the important balance this has for life and survival on earth. The importance of nature and the environment is now interpreted as a web of intricate relationships and its meaning best understood by the phrase *'the web of life'*. Schweitzer's (1929) phrase *'reverence for life'* is often quoted as the ideal conceptual companion supporting and underpinning this notion of a *'web of life'*, which he suggests reverts to the original instruction given to human beings to *'take care'* of the earth in recognition of its majesty.

McKibben (1989) says that currently we *'live in a post-natural world'* and writes about the exploitation and disregard of our environment and our relationship with it. Leopold (1966), a key scholar in the field of environmental ethics, concurs with this position and his construct of a *'land ethic'* encompasses not just the need for a change in the relationship that humans have with the natural world, but also details the way in which human beings must place a high worth on the *'intrinsic value(s)'* of the natural world in which the richness and diversity of our natural environment is paramount and should take precedence in many situations over its *'instrumental'* value, unless preservation and sustainability are present, for example 'man's' harnessing of alternative energies such as wind and solar.

TWO ETHICAL APPROACHES FOR ANALYSIS OF ENVIRONMENTAL ISSUES

The cost-oblivious approach

This approach is rarely practical as it is ordinarily one-sided and best seen as a *'winner takes all'* approach. The use of this approach resonates with duty and rights ethics.

The cost-oblivious approach to environmental conflicts/dilemmas refers to a situation in which the cost of *'repair'* is not considered: the commercial costs are prohibitive as the scope is expansive and often covers entire communities, for example water pollution, mining activity, wildlife conservation/reservation. Such costs are often difficult to uphold in a free-market economy, where costs and profits are important. These costs are invariably met by governments, both national and local, and are related to the budgets they control and the political will extended by the political party in power. In this approach the taxpayer provides the funding solutions. We see the cost-oblivious approach at play in the famous case study in New Mexico in which the quality of the drinking water in the Rio Grande was not of drinking quality and therefore represented a threat to the native American Indians who were taking part in the ceremonies performed in the river. The Albuquerque sewage plant which was responsible for water purification in the Rio Grande could not afford the costs associated with bringing the water in the Rio Grande up to drinking water quality and so taxes were used to ensure the human dignity and human rights of the native American Indians were maintained and their ceremonies could continue. This can be viewed as a cost-oblivious approach in action.

Cost–benefit approach

This approach is extremely practical and is associated with a balancing act in which there is a sense of the *'common good'* that needs to be established and funded. In this approach the cost of *'repair'* is associated with the *'utility'* it will bring and thus resonates with Utilitarianism. However, the adoption of this approach in environmental dilemmas/conflicts runs the risk of marginalisation of the individual and a tendency to ignore special interests. Moreover, often the *'true'* costs are incalculable and accurate assessments of costs and benefits can sometimes include a degree of *'guesswork'*. We can see this in instances such as fracking, quarrying, addressing acid mine drainage and fishing quotas, in which costs are attributed to the activity and the benefits to the communities it represents, versus the costs of not undertaking the activities and the costs in this scenario that will negatively impact the affected communities. In such cases the disadvantages of using this cost–benefit approach are that it can sometimes not consider ethics and ask if what is being proposed is *'right'* or *'wrong'*.

ENVIRONMENTAL ETHICS AND ENGINEERING

Because contemporary engineers work in a social context, they are often confronted by environmental dilemmas and/or conflicts when undertaking engineering projects and activities, especially in South Africa, which is a country rich in national resources and biodiversity. Therefore engineers have a responsibility to use their knowledge and skills to help protect the environment and in addition must be guided by laws (either municipal or national) to assist them in this duty. The ECSA (2017) Code of Conduct details professional expectations for all registered South African engineers when they are developing engineering activities and/or projects that may directly impact the environment in South Africa.

According to section 3.4 of the ECSA code engineers are to;

have due regard for, and in their work avoid or minimise, adverse impact on the environment;

and to;

strive to ensure that in meeting present development needs, the ability of future generations to meet their needs is not compromised.

Whilst not particularly prescriptive, these expectations in the engineering code of conduct nevertheless recognise the importance of the relationship of human beings with the natural world and implicitly would seem to recognise the importance of the *'intrinsic value'* we place upon it now, rather than its *'instrumental value'* of the past.

Arguably ECSA acknowledges the time sensitivity associated with our natural world that has tended to be linear and a focus which is now heavily invested in the future (see Figure 7.1).

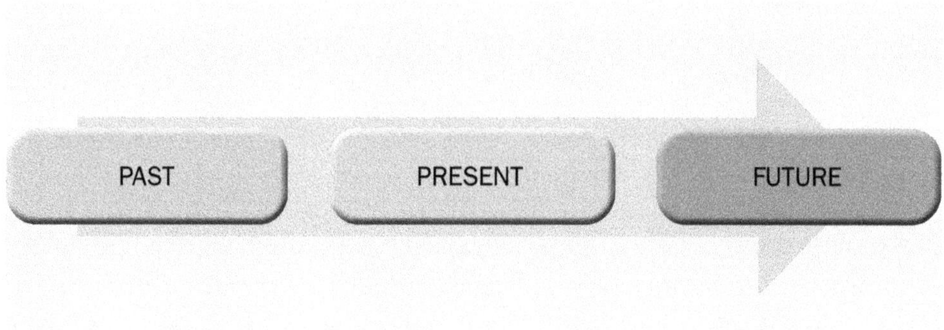

Figure 7.1: The linear nature of our time sensitivity to the environment.
Source: Author's own.

The ECSA code in terms of an engineer's duty to the environment is *'future'*-driven and it seems to respect that our *'present'* is perhaps a function of the *'mistakes'* of our past, when our relationship with our natural world was not one in which we valued the *'web of life'* but one where the *'instrumental'* value of the natural world resulted in a relationship in which there was no consideration for the relationship that future generations could, or should by right, have with their environment.

References

Bible (1952) Revised Standard Version. UK: WM Collins, Sons & Co Ltd.

Blackburn, S (2008) *The Oxford Dictionary of Philosophy.* UK: Oxford University Press, p 116.

ECSA (2017) Code of Conduct for Registered Persons. **https://www.ecsa. co.za/regulation/RegulationDocs/Code_of_Conduct.pdf**

Leopold, A (1966) *Sand County Almanac: With Other Essays on Conservation.* Oxford: Oxford University Press.

McKibben, W (1989) *The End of Nature.* New York: Random House Trade Paperbacks.

Sagoff, M (1993) Population, Nature, and the Environment. *Report from the Institute for Philosophy and Public Policy* 13(4): 10.

Schweitzer, A (1929) *Civilisation and Ethics.* Second edition. UK: CT Campion, pp 246–7.

Singer, P (1993) *Practical Ethics.* Second edition. US: Cambridge University Press, p 268.

Tolkien, JR (2013) *The Lord of The Rings.* China: Harper Collins Publishers.

LEARNING OUTCOMES

- What part has religion played in our understanding of the environment?

- Provide three `buzz' words that are synonymous with our contemporary understanding of the environment.

- Explain the difference between an intrinsic and an instrumental approach to our environment.

- Name and outline a local environmental case study in which a cost–benefit approach has been taken.

- The importance of our environment has been recognised by ECSA. What does it require from its members in this regard?

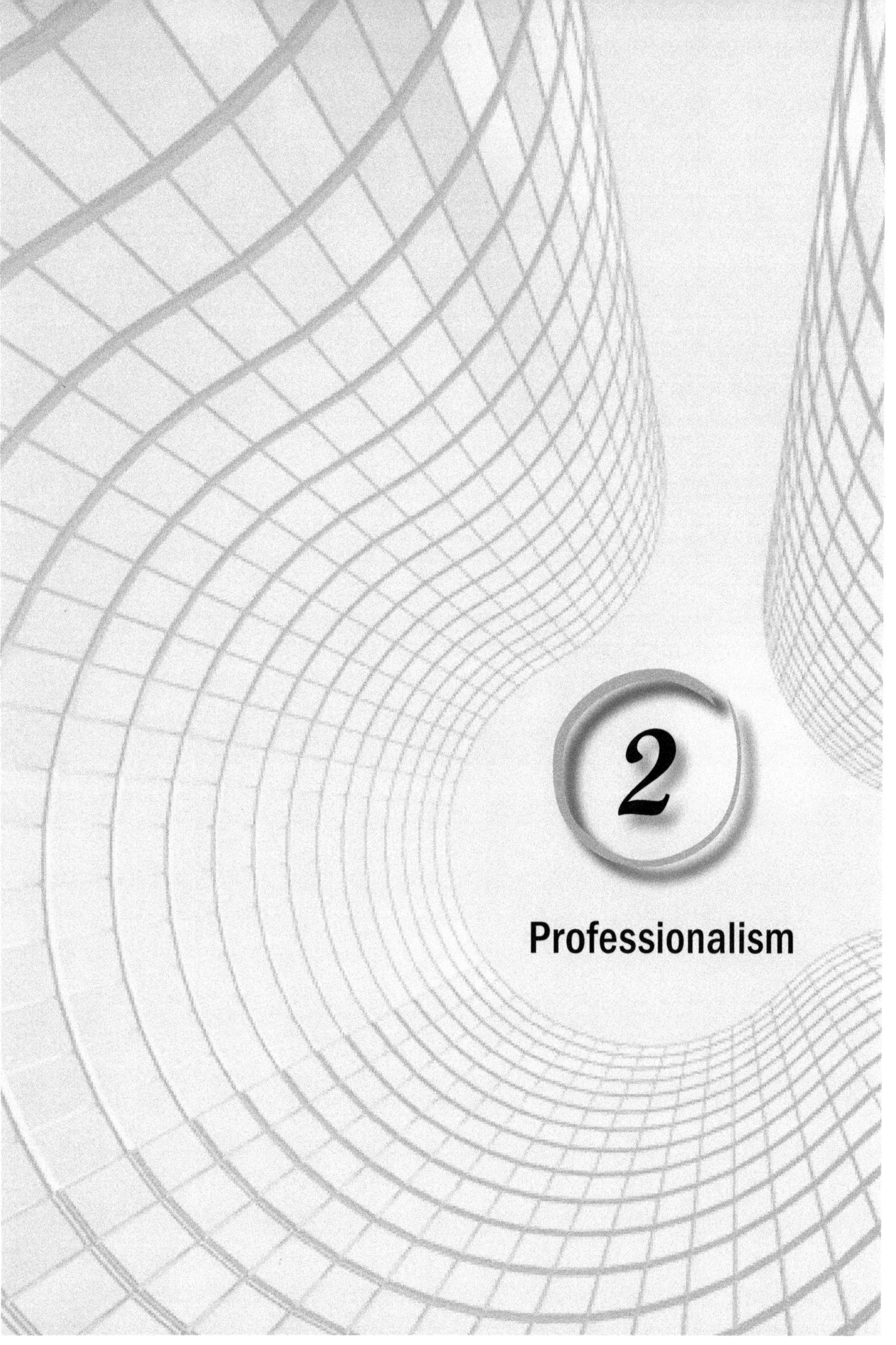

2

Professionalism

PROFESSIONALISM

It is arguably easier for us to define *who* is a professional rather than *what* is professionalism. Both the definition and reality of the term 'professional' has changed significantly over time as forms and types of employment have changed and extended. It is important for us to remember the attributes and characteristics of a contemporary professional engineer do not differ fundamentally from other practising professionals whom we recognise, such as accountants, lawyers, medical practitioners, financial advisors, teachers and the like.

WHAT AND WHO IS A PROFESSIONAL?

An early meaning of the term professional referred to a 'free act' of commitment to a certain way of life. Originally in the West, this was associated with monastic vows whereby a *'public promise'* was given by those entering the monastery and undertaking a life of religious service to both their God and their community. As part of this public promise, suitable candidates were required to 'profess' to be a certain type of person and to occupy a special social and religious role within their community that carried with it strict moral requirements.

By the late seventeenth century, the meaning of professional had become secularised and referred to anyone who 'professed' to be 'duly qualified'.

The *Concise Oxford English Dictionary* (1995) (OED) defines a profession as;

a vocation, calling especially one that involves some branch of learning or science.

However, whilst some agree this is an extremely limited definition for contemporary times, for example one can be a professional model, sportsperson, musician or artist through natural talent rather than being a product of either learning or science; broadly speaking, it could be argued that even natural talent needs to be *'trained'* and honed if it is to become the means of a person being a professional and thus *'duly qualified'*. Perhaps at this point we should be asking ourselves if *'occupation'* is a way to make a living, then . . . how does one make the transition to a profession?

THE DEFINITION OF A MODERN PROFESSION

The American philosopher Mike Davis (1997: 417), who has long been associated with engineering ethics, has provided us with the following definition for a profession which is arguably still relevant. He claims that:

A profession is a number of individuals in the same occupation voluntarily organised to earn a living by openly serving a moral ideal in a morally permissible way beyond what law, market, morality and public opinion would otherwise require.

Davis then expands upon his definition to include the seven characteristics of a profession which he has gathered from both personal observation and research. These are outlined in a book by Harris, Pritchard & Rabins (2009), who share his views that:

1. A profession cannot be composed of only one person. It needs to be composed of a number of individuals.

2. A profession has a public element in that one must openly 'profess' to be an engineer, lawyer, medical doctor or the like.

3. A profession is the way in which individuals earn a living and that occupies them during their working hours.

4. A profession is something an individual can voluntarily enter and exit, providing them with a degree of professional autonomy.

5. A profession has a social component and to that effect must serve the wider community through some morally praiseworthy goal.

6. A profession must serve a morally praiseworthy goal, only through morally permissible means. This precludes professionals from engaging in acts such as deception, coercion and bribery, which are not acceptable as they are considered unethical.

7. A profession is governed by ethical standards with which professionals have a duty to comply. These standards are usually outlined in professional codes of conduct that serve not only the profession but also the expectations of the public as they invariably provide a means of last resort in professional disputes.

It could be argued the philosopher R Max Wideman (1987) has provided us with a condensed list of the characteristics of a profession that draws from the work of Davis (1997) but chooses instead to highlight just five key characteristics of a contemporary profession (see Figure 8.1). These characteristics include professional knowledge, ethics, public service and a professional accreditation body. All such characteristics can be seen to apply to the contemporary profession of engineering both locally and internationally. Most importantly all five characteristics are upheld by their respective professional councils/bodies, referred to by Wideman as a *'sanctioning body'*, with ECSA being no different.

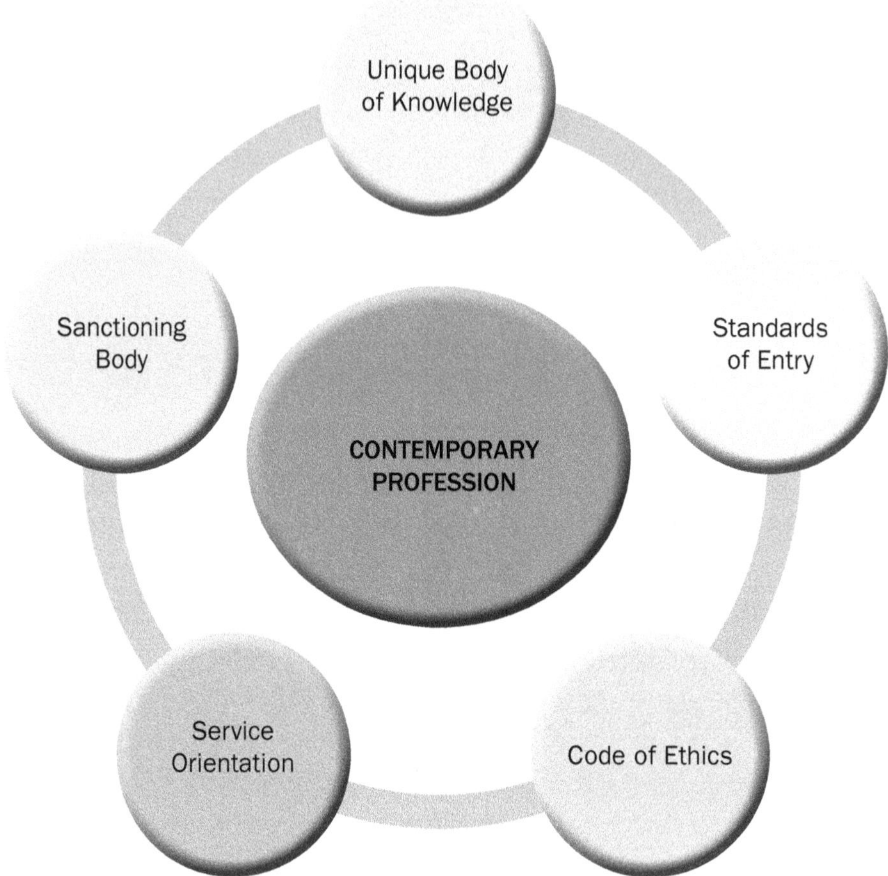

Figure 8.1: The five characteristics of a contemporary profession.
Source: Author's own.

These five characteristics of a profession have been unpacked and scrutinised by scholars to provide us with a universal understanding of these distinguishing and recognised characteristics of a profession.

Within the context of professional engineering, they are adjudged to be as discussed in the sections that follow.

UNIQUE BODY OF KNOWLEDGE

A profession has a theoretical basis acquired through tertiary education, an academy that improves natural talent, or an apprenticeship that focuses on practical skill training. It speaks to extensive training in a specific area of engineering skill.

STANDARDS OF ENTRY

A profession has important knowledge and skills that are vital to the well-being of society. In the case of engineering, the vital knowledge and skills provided by the engineering profession lie in technological advances and public health and safety, and therefore require that professional standards in terms of specific and complementary skill sets are both met and maintained by those entering and remaining in the profession.

SERVICE ORIENTATION AND CONTROL OF SERVICES

Professions invariably operate within a social context. Engineering is no different. Professions also share the characteristic of having a monopoly or considerable control over the provision of their associated goods/services. Therefore a profession has the power to convince a community that only those who have graduated from a professional school should be allowed to hold a professional title. Furthermore, professionalism can persuade society that a licensing system is required and remains the optimal means for both those wanting to enter and practise a profession and for those using their services to verify their professional credentials.

SANCTIONING BODY/CLAIM TO ETHICAL REGULATION (GOVERNANCE AND COMPLIANCE)

Professions have a sanctioning or regulatory body/council that takes the form of offering both professional registration and professional accreditation. It provides a profession with the means of self-regulation. These governing bodies can be wholly independent, or government gazetted and sanctioned and they use the tools of governance and compliance to maintain professional standards and stakeholder expectations. Professional engineering around the world has sanctioning bodies and within southern Africa there is the Engineering Council of South Africa (ECSA) that fulfils this role and function.

CODE OF ETHICS

A code of professional ethics and conduct guides a profession and is best seen as the principal tool used by the profession and its professionals as the means of self-governance and compliance. However, what must never be forgotten is that professional self-regulation is not always a guarantee for ethical conduct amongst its professionals; neither does compliance with an ethical code always mean that abuse is limited. However, a professional code of ethics not only details expected professional conduct but also provides details of professional standards and what the profession considers to be professional malpractice. A professional code of ethics is also the means of dispute resolution between practising professionals and between professionals and stakeholders. Furthermore, a professional code of ethics also provides the means of punishment for registered professionals and can in extreme circumstances prevent them from ever practising their profession again if they have been found to have transgressed both the spirit and principal requirements (otherwise referred to as the letter of the law) of their code of ethics.

THE CHARACTERISTICS THAT DISTINGUISH AND CONTRIBUTE TO THE CONTEMPORARY PROFESSIONAL

Having briefly covered our modern understanding of a profession, it is now time to turn our attention to the modern professional. Scholars and executive coaches seem to concur in the main on the distinguishing features of a profession as outlined above; however, when trying to define the characteristics and core competencies of a professional, interested parties have tried to condense both the rational and behavioural traits into their definitions in an attempt to embrace the challenges associated with the successful practising of a profession.

Mindtools achieves this on their web page (**https://www.mindtools.com/pages/article/professionalism.htm**), where they share their contemporary ideas and their understanding of the characteristics that distinguish the modern professional. They speak of core characteristics and competencies which they believe are required by any contemporary professional, including engineers. They suggest there are six core characteristics that distinguish a professional (see Figure 8.2).

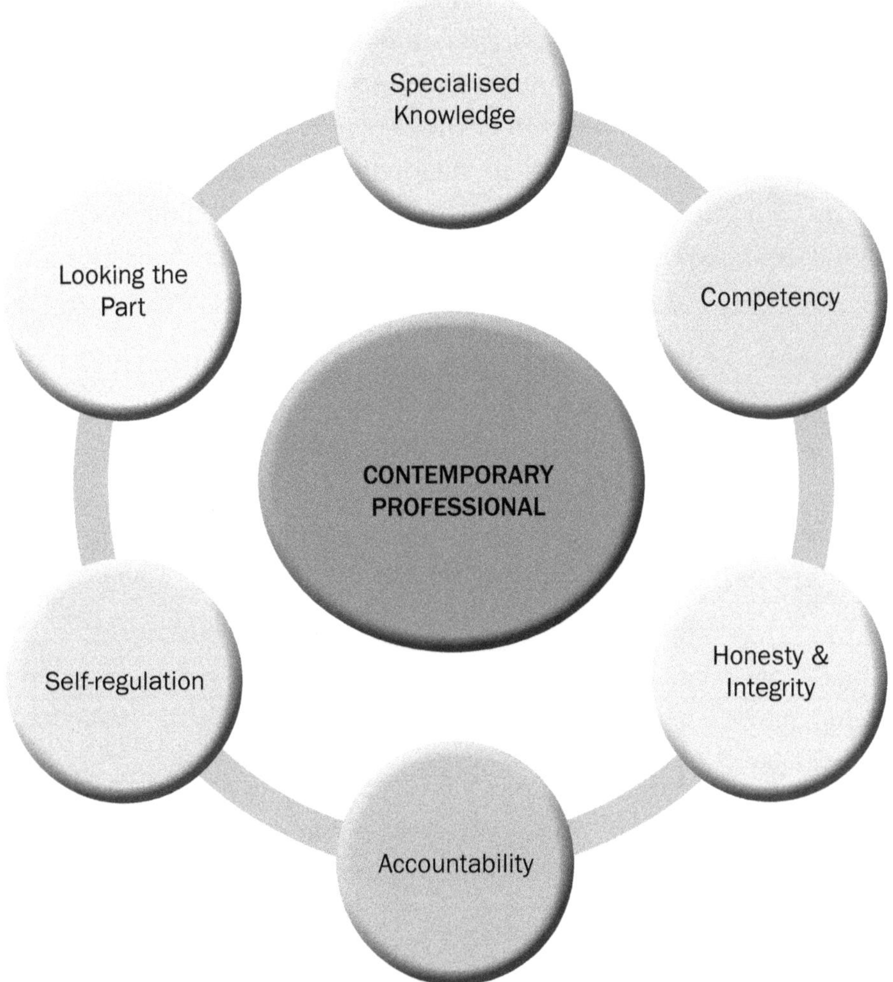

Figure 8.2: The six core characteristics of a contemporary professional.
Source: Author's own.

These six characteristics associated with being a professional can be viewed as a mixture of both rational and behavioural traits and both would seem to have an important part to play in the characteristics and skills that a modern professional must demonstrate. *'Specialised knowledge'* would fall under the former, and *'looking the part'* the latter.

These characteristics of a professional are further endorsed by Preserve Articles **(https://www.preservearticles.com/education/what-are-the-important-characteristics-of-a-profession/18305)**, who has adopted a more ethical stance with regard to the characteristics of contemporary professionalism, for example *'truth and loyalty'* and *'transparency of work'* (see Figure 8.3). However, the further inclusion of *'continuous training'* arguably speaks directly to and acknowledges the dynamic nature of the engineering profession which is at the forefront of cutting-edge technology that can significantly change the nature and type of its engineering activities and engineering projects.

Figure 8.3: The eight core characteristics of a contemporary professional.
Source: Author's own.

At this stage it would perhaps be wise to ask what should take precedence for a contemporary professional – their rational or behavioural characteristics? Moreover, should we be asking what characteristics and skills should be developed by a person who wants to become a successful professional engineer?

RATIONAL AND BEHAVIOURAL CHARACTERISTICS ASSOCIATED WITH SUCCESSFUL PROFESSIONALS

Greg Mason **(https://www.bizhq.co.za/2018/11/23/professionalism/)** is the CEO of bizHQ, a successful South African enterprise focusing on business, executive and leadership coaching. He refers particularly to the behavioural characteristics that need to be associated with professionals and professionalism. He proposes there are ten key characteristics that a successful, modern professional should acquire. Interestingly, they are in the main behavioural group, although he does include some key rational characteristics that should be acquired and demonstrated by contemporary professionals. His ten characteristics are listed below:

1. APPEARANCE: Professionals should dress smartly, neatly and appropriately. They should dress like a winner.

2. DEMEANOUR: Professionals should be confident, polite and warm. They should not be arrogant and should be prepared to listen as well as offer professional advice.

3. COMPETENCE: Professionals should use and apply their qualifications and be seen to be an expert in their field. This is enforced by their qualifications that are on display in their workplace and on business cards and stationery.

4. RELIABILITY: Professionals should meet professional expectations both in terms of their skill set and meeting their deadlines. They must complete the work in which they were engaged in a way that demonstrates both integrity and commitment.

5. ETHICS: Professionals must at all times abide by their code of ethics and behave morally, honestly and ethically within the workplace.

6. RESPONSIBILITY: Professionals must own their mistakes and carry out their work in a structured and organised fashion.

7. MAINTAINING POISE: Professionals must be able to keep calm when under stress, not be caught off-guard and avoid unnecessary confusion and confrontation.

8. PHONE ETIQUETTE: Professionals should always identify themselves using their title, the company they represent and listen carefully to the caller.

9. WRITTEN CORRESPONDENCE: Professionals should ensure their written work is both formal and succinct. It should also form an active part of a paper trail. When dealing with a third party professionals should be polite and provide supporting documentation as and when required.

10. ORGANISATIONAL SKILLS: Professionals must adopt a solutions-based style in the workplace. This demonstrates an uncluttered approach, based on reason, experience and context that meets the expectations of their profession.

Examining Mason's core characteristics of a professional shows they can importantly also be seen to meet and conform to the expectations of those with whom they come into contact, be they colleagues, stakeholders or community members.

DOES ENGINEERING IN SOUTH AFRICA QUALIFY AS A PROFESSION?

Engineering within southern Africa can most certainly be considered a profession. It meets all the criteria and core characteristics for a profession that we have set out above. Its professional body ECSA assists in ensuring this is the case and serves to provide the engineering industry in South Africa with professional engineers who serve in a social context and who comply with both the expectations of the profession and those of concerned stakeholders and the wider community.

References

Davis, M (1997) Is there a Profession of Engineering? *Science and Engineering Ethics* 3(4): 407–28.

Harris, CE, Pritchard, MS & Rabins, MJ (2009) *Engineering Ethics: Concepts & Cases*. Fourth Edition. US: Wadsworth Cengage Learning, p 5.

The Concise Oxford Dictionary (1995) Ninth Edition. Oxford: Clarendon Press.

Wideman, R Max (1987) **http://www.maxwideman.com/papers/spectrum/attributes.htm**

LEARNING OUTCOMES

- List the characteristics of a professional you believe to be the most important.

- Does engineering professionalism in South Africa conform to the definition of a professional? If so, why?

- Explain which characteristic of a modern professional you believe to be the most important.

- Explain the need for behavioural and rational characteristics in contemporary professionalism.

- List five professions in South Africa.

THE ENGINEERING COUNCIL OF SOUTH AFRICA (ECSA)

9

ECSA (**https://www.ecsa.co.za/default.aspx**) is a statutory body established in terms of the Engineering Profession Act (EPA) 46 of 2000 by the democratic government of South Africa. ECSA promotes its purpose as being:

Engineering excellence transforming the nation.

ECSA aims to achieve this vision for the council through its five-fold mission of:

- Determining standards for education and accreditation of educational programmes as well as the registration of engineering practitioners.

- Developing and sustaining a relevant, transformed, competent and internationally recognised engineering profession in South Africa.

- Educating the public on expected engineering quality standards and protecting the interests of the public against substandard quality of engineering work.

- Regulatory efforts to ensure environmental protection through interaction with government agencies.

- Engaging with government to support national priorities.

Like many other professions both nationally and internationally, ECSA provides the means of professional governance for the engineering profession in South Africa. It does this through its Code of Conduct, with which its members are expected to comply. It also provides accreditation of engineering courses at universities both inside and outside of South Africa, namely Mauritius, Namibia and Botswana. ECSA registration is voluntary, but taking responsibility for engineering work requires registration and an acceptance that, by registering with ECSA, a member is agreeing to be bound by its codes of conduct and practice. In line with other professional codes, ECSA registration can be stripped from an ECSA member for breaching the code. From 2025 only a registered engineering professional can take responsibility for engineering work, supervise it and release it.

Therefore, governance and compliance are key to ensuring a consistency of ethics through the provision of an ethical framework in a business, a profession or industry, and the ECSA rules contained within its Code of Conduct are no different.

In South Africa the rules established within this code have been devised and applied with a view to providing the ECSA membership with prescribed guidance on what is considered and accepted professional conduct and professional practice. Once established, this in turn helps to provide the bedrock for the expectations of professional and ethical conduct and practice in engineering in South Africa from both industry and the public.

Governance and compliance are best interpreted as being interdependent and in idiomatic terms represent the two sides of the same coin (see Figure 9.1).

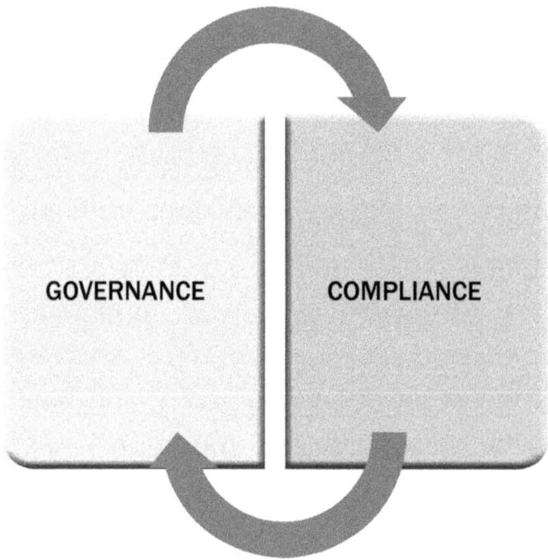

Figure 9.1: The relationship between governance and compliance.
Source: Author's own.

Governance and compliance have a working reality that encapsulates and promotes a successful code of business practice and behaviour (see Figure 9.2). This becomes evident when we examine the roles and functions of ECSA, which are rooted in both its governance of the engineering profession in South Africa and the expected compliance of its engineering membership.

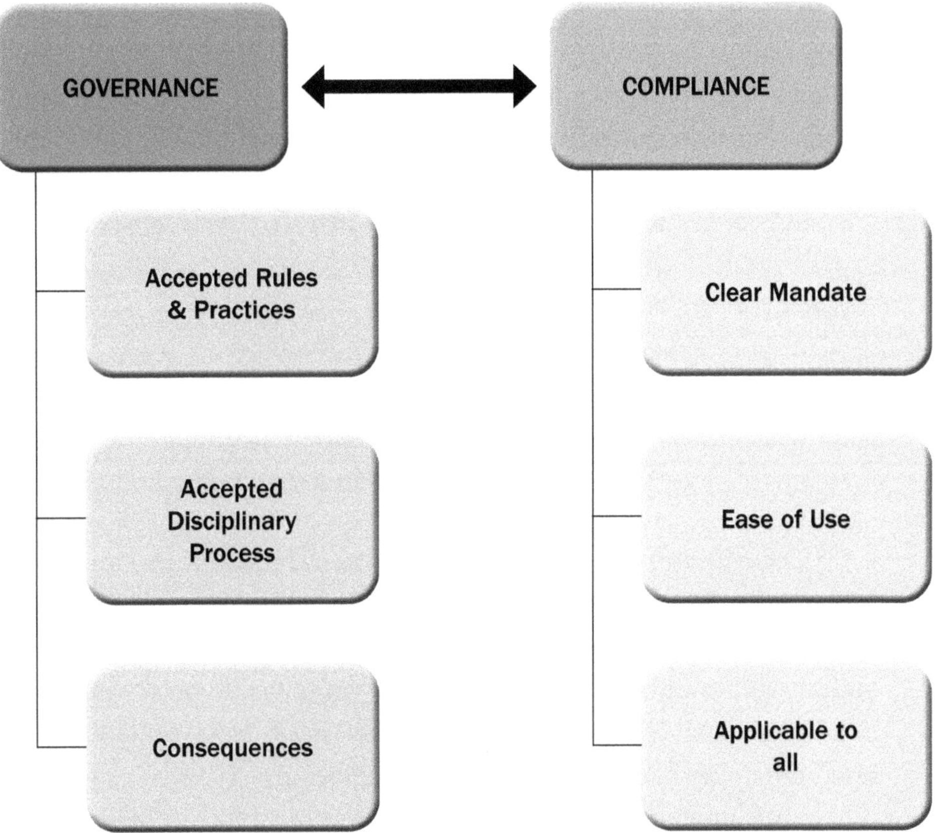

Figure 9.2: The working reality of governance and compliance within the context of ECSA.

Source: Author's own.

As a statutory body, ECSA has key roles and functions that speak to its specific purpose for the profession of engineering in South Africa. These roles and functions also contribute to the way in which the reality of governance and compliance applies in the engineering profession in South Africa.

ECSA'S KEY ROLES

ECSA's key roles are not only focused on the professionalism of engineering as an industry in South Africa but also seek to positively address the expectations of the end users of engineering services.

ECSA's roles fall into four key areas. These are:

1: REGISTRY

ECSA has the role of providing a *registry* for engineers. The ECSA registry provides for both candidate engineers and professional engineers. The former must have an engineering degree that is recognised in terms of the Washington, Sydney and Dublin Accords (**https://www.ieagreements. org/assets/Uploads/Documents/Policy/IEA%20Rules%20and%20 Procedures%20(3%20June%20206).pdf**), whilst the latter must have an engineering degree that is recognised in terms of the same Accords but in addition must have a minimum of three years postgraduate employment in engineering and/or associated industries.

2: ACCREDITATION

ECSA *accreditation* of engineering educational programmes in South Africa is undertaken in consultation with the Council for Higher Education, which is the only legal entity that can approve and accredit higher qualifications. This ensures the qualifications obtained in ECSA-accredited educational programmes meet both national and international standards for professional engineering. ECSA is a signatory of the Washington (1989), Sydney (2001) and Dublin Accords (2002). These Accords gave rise to the International Engineering Alliance, which is in turn responsible for setting standards that provide a framework for international recognition of engineering qualifications that encourage the geographic mobility of professional engineers whilst upholding all signatories to a standard of mutual recognition of engineering education. Therefore, any graduate from an accredited engineering body is expected to be proficient in solving engineering problems at a level that is appropriate to their level of practice, be they an Engineer, Technologist or Technician. In their respective functions they are expected to apply appropriate engineering tools, techniques and methods to design or investigate an engineering problem to the appropriate level.

3: CERTIFICATION

ECSA has a further role which is to provide *certification* by way of professionally certifying engineering technicians, engineering technologists and the like.

4: COUNCIL OF LAST RESORT

ECSA also has the role of providing a Council of *last resort* and as such adjudicates in the complaints process when professional engineering standards and conduct do not meet ECSA's professional requirements.

The key roles of ECSA are the driving force behind its core functions, which can be counted as:

1. **Maintaining stakeholder relationships.** Stakeholders are identified as the triumvirate of industry, tertiary institutions and government.

2. **Constant regulating and monitoring** of the engineering profession in South Africa.

3. **Upholding the ECSA Code of Conduct** among ECSA members (governance), whilst simultaneously ensuring that engineering standards are upheld through industry law(s).

The ECSA Code of Conduct

Like many other professions, ECSA's Code of Conduct for Registered Persons (**https://www.ecsa.co.za/regulation/RegulationDocs/Code_of_Conduct.pdf**) in South Africa incorporates the moral approaches of virtue and duty as tools of governance and expects accredited and registered engineers to comply. The ECSA code is best viewed as protecting and enhancing the lives of professional engineers in their workplace whilst simultaneously providing the same for third parties and the public who are subject to their many and varied engineering activities. In short, the ECSA Code of Conduct assists to uphold both engineering standards and engineering practices in the engineering industry in South Africa.

DETAILS OF THE ECSA CODE OF CONDUCT

The ECSA Code of Conduct consists of six rules that 'govern' the conduct and behaviour of engineers in South Africa and in turn serve the engineering industry as a whole in South Africa. These are:

- **Competency** = *Intellectual virtue*

- **Integrity** = *Moral virtue*

- **Public interest** = *Intellectual virtue*

- **Environment** = *Intellectual virtue*

- **Dignity** = *Moral virtue*

- **Administrative** = *Intellectual virtue*

As can be seen, these six rules of ethical conduct cover the roles and responsibilities of engineering professionals in South Africa and, within the context of moral philosophy, can be interpreted to consist of four intellectual virtues and two moral virtues. As in any professional code of conduct we need to understand their reality. This is the case for both the professional engineer and the engineering profession in South Africa.

THE REALITY OF MORAL VIRTUE IN ECSA'S CODE OF CONDUCT

We have seen earlier in the first section of this book the virtue ethics approach is human-centric and suggests through mentorship we can acquire virtue through habit (moral virtue) and practice (intellectual virtue). We also learned that intellectual virtue is ordinarily measurable, whereas moral virtues are notoriously difficult to 'measure' as they are subject to relativism. With this knowledge, our next question has to be, if examining a professional code of ethical conduct that contains two moral virtues, how can compliance by members of ECSA be judged to have been achieved in the moral virtues expected of them in their professional capacity? How can the acquisition of moral virtues expected of engineers in their professional capacity be seen to be present?

In this case ECSA surmounts this difficulty by creating a universal measure for the two moral virtues in its Code of Conduct, namely **integrity** and **dignity**, whereby it requires its members to *'acquire'* and *'practise'* the following in their regard if they are to be judged to have acquired these moral virtues:

- **Fidelity**

- **Diligence**

- **Efficiency**

- **Professionalism**

- **Standing and reputation**

At first glance it would seem these practices, which have been associated with the two moral virtues within ECSA's Code of Conduct, are themselves subjective and to a large extent semi-formalised, but ECSA has imbued them with a degree of ***'measured relativity'*** within its Code of Conduct for its membership that translates into a working reality for the behaviour and practices expected from engineers in South Africa, who must comply in the following way:

- Fidelity = ECSA members must not engage in any act of dishonesty, corruption or bribery. They must also not divulge any information of a confidential nature which they have obtained in the exercise of their engineering practice and activities.

- Diligence = ECSA members must ensure that any work approved or certified by them has been reviewed or inspected to the extent necessary to confirm the correctness/accuracy of the approval or certification. They must also not misrepresent or knowingly misrepresent their own or any other person's academic or professional qualifications or competency.

- Efficiency = ECSA members must discharge their duties to their employers, clients, associates and the public with integrity, fidelity and honesty.

- Professionalism = ECSA members must actively avoid situations that give rise to a conflict of interest(s) or have the potential for such conflict of interest(s).

- Standing and reputation = ECSA requires that members may neither personally nor through any other person improperly seek to obtain work, or by way of commission or otherwise, make or offer to make payment to a client or prospective client for obtaining such work.

As can be seen, the first and the last 'relative measures' are interrelated and confirm that 'fidelity' or 'trust' is essential and measurable in the quest for maintaining an acceptable standing and reputation for the engineering profession in South Africa and those professional engineers who are registered members of ECSA. Moreover, it also provides stakeholders and the public with clear expectations of the conduct and behaviour that should be demonstrated by any engineer registered with ECSA and practising their engineering skills within South Africa. It also serves to provide a benchmark for the behaviour and practices of those in the engineering industry in South Africa which, if not met, can be challenged.

THE REALITY OF INTELLECTUAL VIRTUE IN ECSA'S CODE OF CONDUCT

The remaining four rules found in the ECSA Code of Conduct have been identified previously as intellectual virtues. They are:

- **Competency**
- **Public interest**
- **Environment**
- **Administrative**

Interestingly, these virtues can also be interpreted as duties/responsibilities/ obligations that must be undertaken by registered engineers who are members of ECSA. To this end, the rule of **public interest** can be viewed as a *'categorical imperative'* for any ECSA-registered engineer as it rules that priority must be given by engineers to public health, safety and interest at all times. In other words, an engineer's duty must be *always to prioritise public health, safety and interest.*

Other intellectual virtues cited in the ECSA Code of Conduct, namely **competency, environment** and **administrative,** can also be interpreted as duties of the code. However, these duties fall within the realm of 'hypothetical imperatives' as their interpretation can be contingent and they contain the word 'if'. The remaining intellectual virtues cited in the code that can equally be seen to conform to duties are:

- Competency = It is the duty of ECSA members to notify ECSA if an engineering professional is declared medically unfit. It is also their 'duty' to declare criminality if they want to register with ECSA. In addition, professional engineers are not registered in a branch of engineering, but can practise in any discipline provided they take responsibility for developing competence in this area. Therefore a professional engineer is unable to undertake work outside of their area of competence.

- Environment = It is the duty of engineers to avoid or minimise any adverse impact on the environment. It is also the duty of engineers to adhere to generally accepted principles of sustainable development. Within a South African context there is an *'additional duty'* attached to the rule of the environment for engineers and engineering practices of mindfulness of any cultural claims on the environment, for example ancestral burial sites or battlefields. It can be seen as their duty to halt engineering activities if an ancestral claim to the environment is discovered until public engagement has resulted in the engineering activity being approved by the community.

- Administrative = It is the duty of engineers in terms of engineering designs and engineering activities to ensure that all and any work information is issued, signed and dated by a registered professional engineer. It is also the duty of engineers not to place contracts or orders if the client has not provided written authority. It is also the engineer's duty not to destroy or dispose of, or let anyone else dispose of, any information within a period of ten years after completion of the engineering work/activity.

CONFLICT OF VIRTUE AND/OR DUTY WITHIN A PROFESSIONAL CODE

It has been observed that despite the governance that professional codes of ethics and conduct seek to provide, the glaring omission or obfuscation of guidance they provide for situations in which there might be a conflict of virtue and/or duty for their members is apparent. Such codes rarely provide a hierarchy of virtue or duty for situations when there is a conflict in behaviour and/or practice for a professional. Perhaps in this instance ECSA does have the benefit of citing the categorical imperative of 'public interest' for its registered members and the engineering industry in general within South Africa and the understanding this must take precedence in any engineering activity undertaken.

Professional codes of conduct can also lack specificity or prescription and here ECSA is no different. Its Code of Conduct requires members to *'keep records for ten years after completion'*, but it does not prescribe to its membership how this must be done. Must records be kept in a password-protected computer? Must records be kept in a fire-proof cabinet? Must records be stored in files in 'the cloud'? In such instances, we must hope engineering professionals will take it upon themselves to ensure that confidentiality, trust and document security are maintained at all times.

LEARNING OUTCOMES

- Discuss the need for a professional code of conduct and practice.

- What are the roles and functions of ECSA?

- What are the weaknesses of ECSA's Code of Conduct?

- Can the six rules be seen as conforming to the virtue approach, the deontology approach or both?

- Which rule is considered a categorical imperative and why?

PUBLIC ENGAGEMENT

10

Around the world, the work and conduct of professional engineers is overseen by engineering councils who proactively promote a professional code of conduct that has both internal and external benefits for members and the public. As we have seen, it serves to affirm and uphold professional expectations whilst implicitly being a channel through which trust can be built between engineering professionals and the communities and individuals they serve. Nowhere is this trust more evident than in the area of public engagement.

Contemporary engineers, unlike their predecessors, are expected to acquire the communication skills associated with public engagement because their professional work is within a social context. This has led to the need for professional engineers to acquire communication skills that can be applied during public engagement forums to inform those affected and/ or impacted by their engineering activities with the objective of consensus being reached by all attendees (see Figure 10.1).

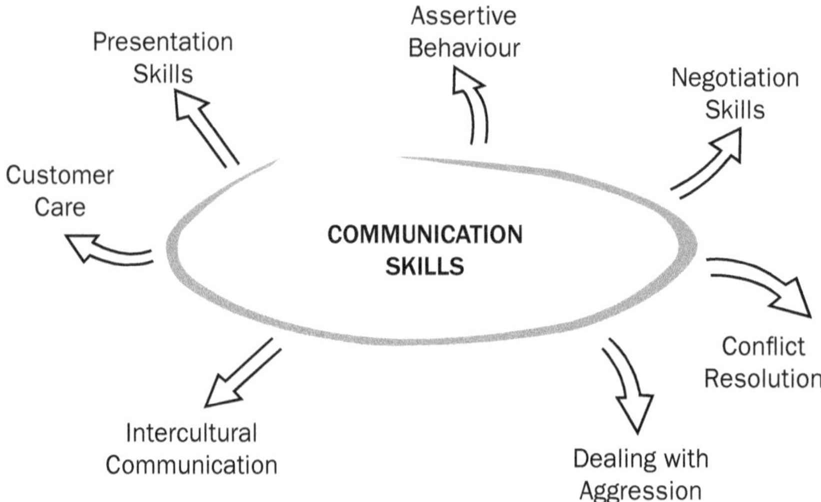

Figure 10.1: The skills required for successful communication by engineers in a public forum.

Credit: Communications Skills stock photo by ©vaeenma 73966315.

Contemporary engineering activities can often affect both the individual and the community and need to be explained in a public forum to ensure that deadlines, expert witness statements, impact/risk analysis, mitigation and project goals are shared. To successfully meet such undertakings, engineers need to have a keen sense of their responsibilities/duties to their profession and the public alongside an equivalent awareness of what constitutes human dignity.

TYPES OF PUBLIC ENGAGEMENT

Essentially, public engagement can take two forms and can either be seen to be *proactive* or *retrospective*.

Proactive

This is public engagement that takes place before a proposed engineering project. This is the preferred method.

Restrospective

This is public engagement that takes place **during** an engineering project or **after** project completion. This is not an ideal situation for either the engineer(s) or the community as both parties feel sidelined and the project can sometimes be seen as a 'runaway train', with neither representatives from the engineering profession nor the community able to impede its progress towards its ultimate destination.

THE PURPOSE OF PUBLIC ENGAGEMENT

As mentioned earlier, there is a need for trust to be at the centre of all public engagements if their outcomes are to be the result of engineers, individuals and communities coming together. To achieve this, there is a need to engage with one another and undertake robust discussions with the goal of reaching consensus or at least compromise on engineering projects that impact a community.

In their book *Engineering Ethics* Harris, Pritchard & Rabins (2009) make it quite clear that trust needs to be both demonstrated by engineers when undertaking public engagement as well as earned. To ensure this, there needs to be effective communication by the engineer(s) to/with the community and vice versa.

Usually, public engagement takes the form of a public meeting in which the agenda is not for 'information only' but serves the purpose of a forum in which the exchange of information and ideas from interested parties is directed and from which views can be heard and action can be taken. This public forum should provide the space and the opportunity in which professional engineering and community concerns can be shared and addressed. Such an approach leads to public meetings being called with the aim to serve the following objectives:

- To further the democratic process in community decision-making. Those community members in attendance should not feel intimidated in terms of not being able to express their views. They should rather feel themselves to be in a *'safe space'* where they are able to *'tell it as they see it'*.

- To provide a neutral ground for interested and affected parties to air their views and grievances openly as well as the benefits of the engineering project/activity.

- To provide an agenda that can embrace all arguments and concerns surrounding an engineering project/activity and be convened in such a way that differences in opinion can be aired clearly and concisely without violence and with appropriate supporting data.

- To provide a platform for social justice in that public engagement should be the means of *'levelling the playing field'*, in which expert witnesses can speak on behalf of the professional engineers. Members of the community can also engage their own expert witness to repudiate as well as confirm the facts shared by their counterparts representing and supporting the engineer's viewpoint.

Whilst such public meetings/forums can often be accompanied by extreme emotion, accusation, ignorance, antagonism, technical data and specialist knowledge, they should never be used as a means of obfuscation; they should rather be used as a means of elucidation and dialogue for all parties concerned because, in large part, they can be seen to be a forum where opposing parties can meet and negotiate in the best interests of all concerned.

The following situations, should they occur, can be seen to **prevent** this from happening;

- Manipulation

- Overtly violent protest

- A *'might vs right'* approach to the issue under review, as this often marginalises human dignity

- A *'short-term gain, long-term pain'* approach that experience suggests does not provide any engineering project in a community with full acceptance

Those in attendance at public meetings/forums seeking to discuss the impact of engineering projects/activities, can be drawn from individuals occupying any of the following roles:

- a lobbyist

- an activist

- an expert witness

- a politician

- a lawyer

- a professional engineer

- a community leader/member

- a tribal elder

- a financier

- an environmentalist

These people would be required to not only bring their experience and qualifications to bear in the public meeting but also to speak on behalf of those whom they represent, by engaging with those present in a way that helps to position and quantify their concerns, reservations, outright objection(s) or support for the engineering project/activities under discussion.

THE PRINCIPAL PREREQUISITE FOR SUCCESSFUL PUBLIC ENGAGEMENT IN A PUBLIC MEETING

Honesty

For trust to occupy its rightful place in public engagement, the honesty of all parties is a critical prerequisite. The importance of honesty and 'truth-telling' in modern engineering practice particularly refers to the areas of:

- professional judgement

- communication

Both skills can be seen to relate directly to the ECSA Board Notice 41 of 2017 (**https://www.ecsa.co.za/regulation/RegulationDocs/Code_of_Conduct. pdf**), in which ECSA details its expectations of how registered engineers should conduct themselves. The Code of Conduct for Registered Persons that forms part of the Engineering Profession Act 46 of 2000 specifically states in section 3.2 – Integrity of registered persons – that engineers;

(a) must discharge their duties to their employers, clients, associates and the public with integrity, fidelity and honesty.

ECSA goes on to state that under no circumstances must registered members;

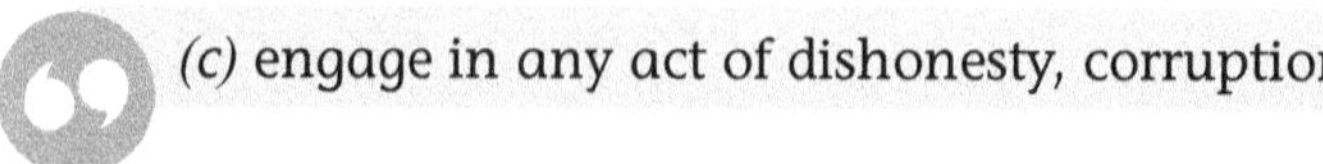

(c) engage in any act of dishonesty, corruption, or bribery.

In addition, ECSA is unequivocal that its registered members;

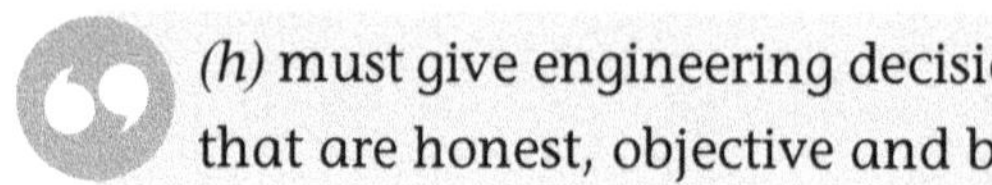

(h) must give engineering decisions, recommendations or opinions that are honest, objective and based on facts.

CHARACTERISTICS THAT PREVENT THE SUCCESS OF PUBLIC ENGAGEMENT IN A PUBLIC MEETING

Lying

As we have seen, honesty is critical for successful public engagement and is a requirement of registered engineers from their engineering councils. It should then come as no surprise that lying can derail a public meeting and lead to a lack of trust between those engaged in the public forum. Harris, Pritchard & Rabins (2009) suggest that dishonesty can take the following three distinct forms, all of which, either singularly or collectively, should be avoided at all costs. These are:

- the withholding of information

- failure to seek the truth

- deliberate deception

Withholding of information

This is deceptive behaviour and is undertaken to ensure the 'whole picture' is not shared in the public forum/meeting. It can alternatively be called *'lying by omission'*. This has been used quite successfully in the context of environmental ethics, when the full details of the effects of engineering activities in South Africa, such as fracking and mining, are not shared with a community by engineers and expert witnesses, for fear it could derail the engineering project/activities and its profits.

Failure to seek the truth

An example of this would be a case in which engineers commissioned and received carbon emission results for a planned light industrial park they were hoping to commission in a bid to create local employment. The results provided raw data for the project over a single month in summer. However, the engineers are aware that further tests could/should be done if the projected annual results of carbon emissions are to be accurate and include the weather variations that are prevalent in autumn when high winds are typically experienced and would negatively affect the results of carbon emissions. In such a case, failure to undertake these additional carbon emission tests in autumn could be viewed as a failure by the engineer to *'seek out the truth'* as the *'true'* annual expected carbon emissions should not be a function of a monthly emission test taken in a single summer month and multiplied by twelve. It should rather include twelve monthly tests over the course of a year to present an accurate representation of the expected annual carbon emissions for the light industrial park. These accurate results would then form a major part of the information shared during the public engagement forum/meeting.

Deliberate deception

This is perhaps considered to be the worst type of dishonesty as it implies premeditation and can have serious consequences in a public forum/ meeting. Examples of deliberate deception can range from those actively taking part in public engagement who falsify their credentials and/or expertise, to those who misrepresent findings, choosing to present only results that support their position, to those who only report on key income streams on a balance sheet whilst choosing not to report unidentifiable costs that negatively impact profits that were scheduled, for example, to form part of a community upliftment project.

Conflicts of interest

This can also serve to destroy trust in public engagement between engineers and the community. An example of this would be if there was a hydropower engineering project on the public engagement agenda, and the engineer assigned to lead the project was a significant shareholder in the company recommended to undertake the supply of turbines. Their interest in the engineering project could be seen to be over and above its merits, as their interests lie in the profitability of the turbine company both as a major shareholder and as a lead project engineer. In short, their position within the public forum would be severely compromised.

References

Harris, CE, Pritchard, MS & Rabins, MJ (2009) *Engineering Ethics. Concepts & Cases*. Fourth Edition. Canada: Wadsworth, Cengage Learning, pp 190–1, 216.

LEARNING OUTCOMES

- What is the preferred type of public engagement?

- Provide three examples of people who would attend a public forum to discuss engineering activity.

- What is the principal prerequisite for successful public engagement and why?

- List the three types of lying that can occur in a public forum.

- Does the ECSA Code of Ethics have professional expectations of its members when engaging in a public forum? If so, what are they?

RISK, SAFETY AND MITIGATION

11

All three of these terms – risk, safety and mitigation – are of critical importance to engineers and their profession. Their significance takes on a further dimension when we remember their synthesis becomes the most important duty of an engineer whenever and wherever they practise their profession, which is to protect the safety and well-being of the public.

This paramount duty (or as we have learned earlier in Kantian ethics, an engineer's categorical imperative) to protect the safety and well-being of the public, appears to be a global requirement and is featured worldwide in all engineering codes of ethics.

To this end, ECSA is no exception and specifically cites this duty in sections 3.3(a) and 3.3(b) under its 'Rules of Conduct: Ethics' **(https://www.ecsa. co.za/regulation/RegulationDocs/Code_of_Conduct.pdf)** as Public Interest.

ECSA states in section 3.3(a) that registered persons;

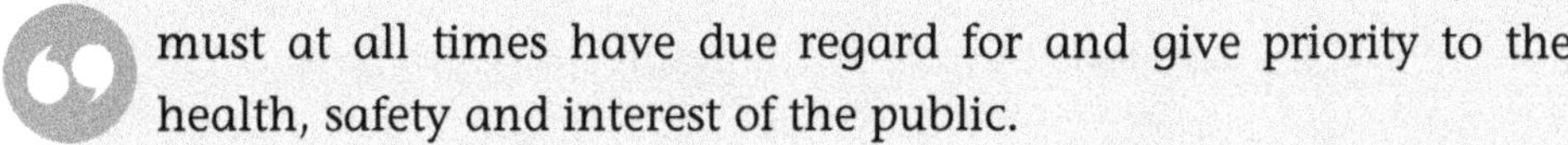

> must at all times have due regard for and give priority to the health, safety and interest of the public.

ECSA extends this duty for registered engineers further in section 3.3(b), when it further tasks them with the directive they;

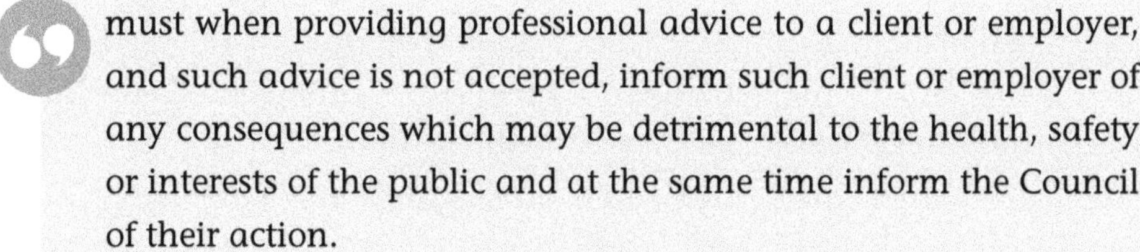

> must when providing professional advice to a client or employer, and such advice is not accepted, inform such client or employer of any consequences which may be detrimental to the health, safety or interests of the public and at the same time inform the Council of their action.

However, when these three concepts – risk, safety and mitigation – are interpreted individually, we can begin to see how the sum of the parts in this case are as important as the whole for contemporary engineering professionals in South Africa.

SAFETY

The concept of 'safety' can be vague because it can be a value judgement. In other words, it is qualitative or can be seen as subjective and/or relative, for example swimming in an estuary if you are a strong swimmer could be considered 'safe', but is it safe on a day when the tide is coming in and the currents are strong? Similarly, it is expected that bridges are a 'safe' structure designed by engineers to span and carry loads and that will not collapse when traversed. However, experience can provide us with evidence to the contrary. Similarly, it is expected that building constructions once completed are permanent and will not be subject to weather conditions, design flaws, wind speeds and directions, or inadequate construction techniques; but once again, there are numerous engineering case studies that can provide us with information and evidence to the contrary.

It has been considered that if something is *'safe'* the corollary is that it is without *'risk'*. However, life experience tells us otherwise and suggests there has to be a quantifiable balance between the two states of *'risk'* and *'safety'*. It is at their confluence that professional engineers are often required to arbitrate using their considerable knowledge, skills and experience in a way that recognises and acknowledges that any risks that might be present must be mitigated through the use of an engineering solution. Once this is undertaken, the situation has been *'made safe'* through the professional intervention of an engineer(s) and their ultimate professional duty to 'public interest' has been carried out.

The ISO 31000 Risk Management Process (see Figure 11.1) provides us with a graphic explanation using the seven steps that need to be taken by an engineer when identifying risk and mitigation in an engineering activity and/or engineering project if safety is to be the goal.

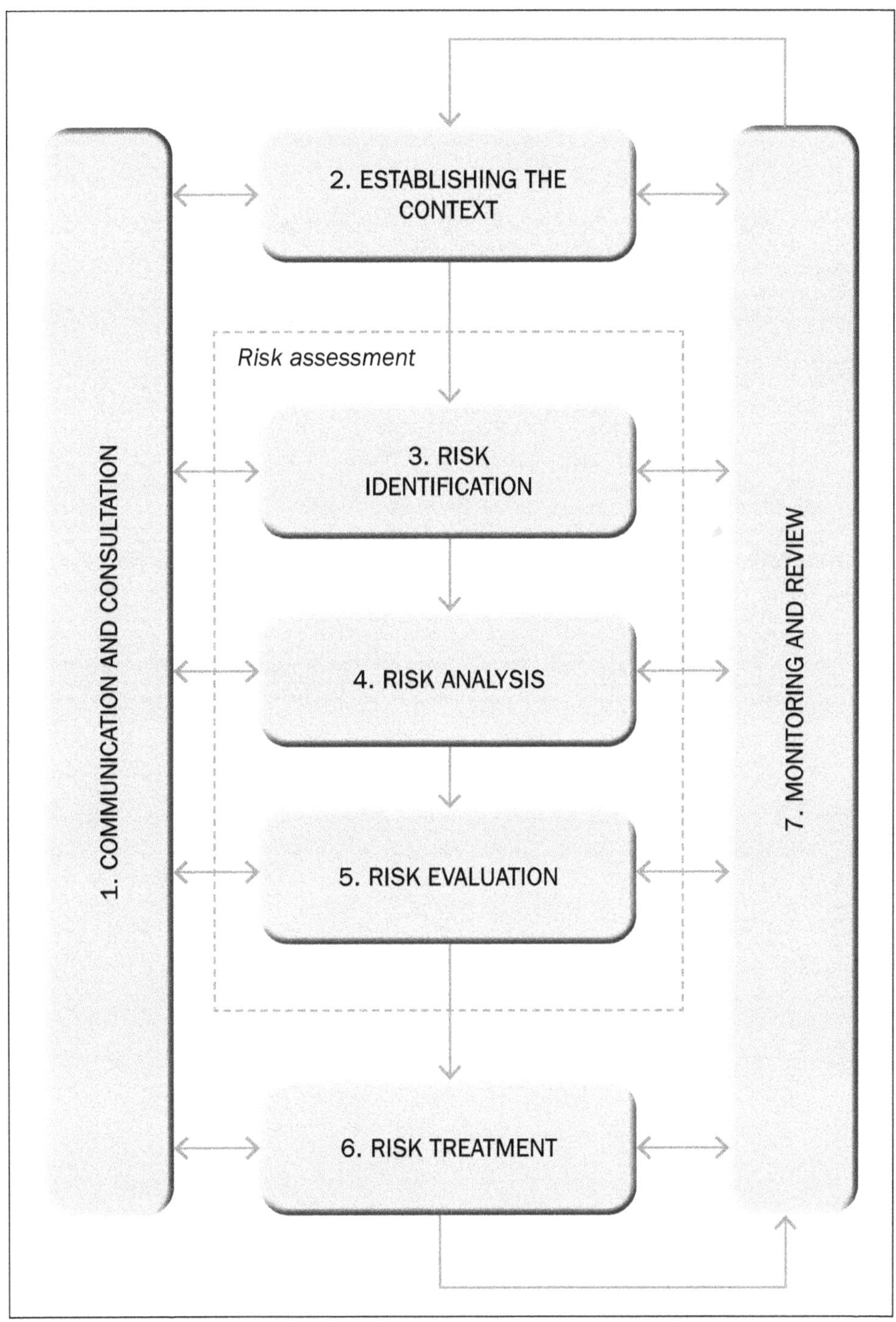

Figure 11.1: Risk management process.
Source: https://www.researchgate.net/publication/267488928_Risk_Management_
in_Product_Design_Current_State_Conceptual_Model_and_Future_Research/
figures?lo=1

All seven steps of this process are equally important for the engineer and each is dependent on the correct completion of its predecessor if mitigation and safety is to be the end result of the process.

The list below details some of the key aspects that can arise in engineering activities and/or engineering projects and how they can sometimes become both the cause of risk and its means of mitigation for contemporary engineers:

- **Costs** (Attributable in terms of financial, physical, community, environmental and reputational.)

- **Project deadlines** (Are the project timings realistic or unachievable?)

- **Labour** (Is there a chance of unrest? Is there evidence of use of unsuitable/inadequate skills for the task in hand?)

- **Materials** (What and how are materials to be used? Is their integrity assured? Do they meet accepted safety standards?)

- **Community involvement** (Does the public accept and/or oppose the engineering project/activity as a whole, or just certain parts of it?)

Each one of these key fields can undergo risk assessment by the engineer, via the risk management process if they present as compromising safety in an engineering project or activity and a cause for risk that requires mitigation.

SOUTH AFRICAN BUREAU OF STANDARDS (SABS)

According to the overview on its website (**https://nationalgovernment. co.za/units/view/161/south-african-bureau-of-standards-sabs**):

> The SABS is mandated to: develop, promote and maintain South African National Standards (SANS); promote quality in connection with commodities, products and services; and render conformity assessment services and assist in matters connected therewith.

Throughout the world, countries have recognised governmental/national safety bureaus that develop, promote and enforce safety standards across a range of industries and products. Whilst these standards can be accused of only enforcing the minimum safety requirements, they nevertheless serve both industry and the public with an accepted system that offers the benefits of impartial governance and the absolute need for compliance recognised in law. Should an engineer undertaking engineering activities in South Africa not comply with these standards and an accident or disaster occur, one of the first aspects a forensic team will investigate is whether accepted SABS products and practices were used. Non-compliance with these standards serves as an indication that safety was not a priority for those involved and the full force of the law will be brought to bear on the situation. The accused may not only have professional licences withdrawn but also be liable to pay fines and possibly serve time in prison too.

INSURANCE AND RISK MITIGATION

Whilst the risk management process identifies the steps an engineer must take for successful risk assessment, it does not negate an *'objective measure'* of risk and mitigation for engineers to use with a universal application which can take the form of insurance. For this to be a working reality, the contemporary professional engineer has to not only correctly and accurately assess the risk personally, but also needs to achieve this in concert with the expertise of actuaries skilled in the fields of understanding and measuring risk profile, and the financial compensation if risk mitigation is to be both acknowledged and paid in times of adversity and/or disaster.

The types of insurance that engineers could explore as a means of risk mitigation in engineering activities and/or engineering projects are discussed below.

Voluntary vs Involuntary

Sometimes risk acceptance can be a function of knowledge.

Voluntary

People often consider something is safe/safer if they knowingly (voluntarily) accept the risk, for example 'I will make this bungee jump because I will have a body harness'; 'I will choose to buy the cheaper property next to the cell phone tower'. This scenario can also play out in engineering activities/projects.

Involuntary

People would not accept the risk had they prior/full knowledge of the situation, for example 'I would not have bought this property if I knew the high levels of radon gas that were being emitted'; 'I would not have undertaken this bungee jump had I known the harness was twenty years old and was made from inadequate materials'. Such a scenario can also play out in engineering activities/projects as well as bedevil ever-present engineering deadlines.

Short term vs Long term

Sometimes something with negative short-term consequences seems to skirt around the edges of 'risk' when compared with the long-term consequences. The following is an example of how this can present itself in the context of an engineering activity.

Short term

'I will use the traditional materials and methods previously associated with this engineering activity, as the existing emissions tests and industry standards are met.'

Long term

'Whilst existing emissions tests provide legal standards, there is some concern regarding their accuracy as they do not include recent climate change findings. In time, new standards will be imposed to mitigate these concerns that point not only to severe health issues but also to accompanying climate and social issues if new legal emission standards are not followed soon. The engineering project will consider this.'

Expected probability

This refers to *'an educated guess'*, for example a 100-year high-water mark measure is often used by engineers for their engineering activities, including site constructions, in the belief that if precisely followed, the probability of property flooding has been mitigated and will not present risk of injury and loss of life to those in the community. However, this is not to say the unexpected cannot occur, despite strict adherence by engineers to these measurements, for example Hurricane Katrina (2005) and the Indian Ocean Tsunami (2004).

The former was unexpected and described as *'one of kind'* that not only exposed the contemporary dangers of constructing dwellings below sea level, but also the inadequacy of the levee system which was used to mitigate the risk of flooding in adverse weather conditions in the area.

Reversible effects

Often engineering projects and/or activities seemingly carry 'less risk' if the attendant risks/consequences are ultimately reversible, for example we will design and build a bush dwelling with a thatched roof because, should excessive wildlife activity begin to destroy the thatch, we can clad the dwelling with roof tiles. Another example of a reversible effect would be one in which engineers use a bolt connection for a beam in their construction project, knowing that if it shows signs of distress, they can and will replace it with a rivet.

Threshold levels

If there is considered to be a threshold for the effects/consequences of something, then there will be a greater tolerance for the associated risks, for example the chance of proprietary software not being able to load up on a company laptop; the chance of a catalytic convertor failing in the first 5 000 kilometres driven in a new car.

Delayed vs Immediate

This relates to activity in which harm or negative consequences are delayed and in so doing their use can be deemed acceptable, for example the burning of fossil fuels is the greatest contributor to global warming, but it will take 100 years (four generations) for the polar ice cap to melt if their use continues on its current scale; there are estimated to be more than 6 000 satellites orbiting earth in 2022 of which the number of active satellites is only just over 3 000. The balance is considered space debris and there is no commercial incentive to act against space debris as polluters are not assigned costs, so space debris will continue to grow.

THE PROFESSIONAL ENGINEER'S DILEMMA

This can be seen to be three-fold and is explained below.

Accurate assessment of risk profile

Risk assessment and risk mitigation should be undertaken on a project-by-project basis. It is rare that any two projects are the same, as context and time are always variable. However, there will be a need for every engineering professional to comply with SABS.

The need for continuous professional development (CPD)

Professional engineers must stay abreast of 'new' findings, practices, materials and standards that affect risk profile and assessment and accepted safety standards and protocols. ECSA provides this platform for its registered members and encourages CPD as a means for those involved in the engineering industry in South Africa to have contemporary knowledge that can be utilised across a variety of engineering activities.

The need for ethical/moral judgement

Engineers should in addition exercise both ethical and moral judgement. They should not use purely economic, personal or political requirements when judging risk, mitigation and safety. Every engineer has moral agency which they should exercise when required (particularly in the pursuit of human dignity, which can be ignored or at best neglected in the risk assessment and risk profile process used by professional engineers) and not forget or neglect their engineering categorical imperative in terms of public health and safety.

LEARNING OUTCOMES

- Compare risk and safety. Can you have one without the other?

- Risk assessment and mitigation are key in engineering project management. List five types of risk mitigation.

- Explain how unassessed risk can jeopardise an engineering activity and/or project.

- Why are public health and safety important for the modern professional engineer?

WHISTLE-BLOWING

Whistle-blowing is usually associated with situations where a moral dilemma/conflict occurs resulting in a conflict between 'rights' and 'duties', and within a professional context it is often highlighted when disregard of a code of conduct and/or corporate policy occurs. It is ordinarily undertaken by an individual but can, in certain instances, be undertaken by a professional body or an organisation. A useful definition of whistle-blowing that I have developed and like to use is:

> A response made to draw attention to unethical or illegal activity and/or behaviour.

ORGANISATIONAL TYPES OF WHISTLE-BLOWING

Fleddermann (2012) reminds us that whistle-blowing can take two forms and can be considered in the capacity of either internal whistle-blowing or external whistle-blowing. The differences between these two forms can be seen below.

Internal whistle-blowing

This occurs when the act of whistle-blowing is kept within the company or organisation. It can be viewed by the company or organisation as an act of disloyalty by the employee or a member towards not just the company or organisation but also their fellow employees and colleagues.

External whistle-blowing

This occurs when the act of whistle-blowing extends beyond the company or organisation and use is made of social media, the press, TV, radio, police, government, government agencies, law and professional bodies.

PERSONAL WHISTLE-BLOWING: THE WHISTLE-BLOWER

Personal whistle-blowing refers to a person who undertakes whistle-blowing. They are called a whistle-blower and in deciding to undertake this activity, they can choose one of two routes. Whistle-blowers can choose to be either an acknowledged whistle-blower or an anonymous whistle-blower. The differences between these two types of whistle-blower are explained below:

Acknowledged whistle-blowers

Situations where individuals make themselves known, in other words they put their name to an accusation of unethical/illegal activity(ies). They see whistle-blowing as their *'personal duty'* and as the only course of action open to them in order to get resolution.

Anonymous whistle-blowers

Situations where accusations of unethical/illegal activity(ies) are made with no reference to the source.

WHEN SHOULD YOU BLOW THE WHISTLE?

Whistle-blowing is undertaken by those with a refined sense of duty and virtue and a well-developed sense of human dignity. Whistle-blowing invariably refers to a situation in which an assessment has been made by a whistle-blower that uncovers a total disregard for *'human rights'* and by extension *'human dignity'* or alternatively the situation can be the result of a neglected professional duty or, in extreme circumstances, both. Thus, whistle-blowers have either directly or indirectly uncovered a situation that is either illegal and/or unethical and they seek to bring the details into the public domain with the aim of drawing attention to the matter in a way that will start a process of investigation and resolution. To this end, it has often been suggested that whistle-blowers seek justice.

Fledderman (2012) suggests that for whistle-blowing to be successful, four conditions must be met: need, proximity, capability and last resort.

Condition 1: Need

A sense of proportion must accompany the whistle-blower's interpretation of the unethical or illegal activity(ies). They must ask themselves: 'Is the activity worthy of further investigation?'

Condition 2: Proximity

The whistle-blower must have *first-hand* knowledge of the unethical and/or illegal activity(ies). The word of a third party is insufficient.

Condition 3: Capability

The whistle-blower must correctly *'gauge the likely success'* of their action. Are there recognised processes and procedures that need to be followed to register the illegal and/or unethical activity identified by the whistle-blower? Has the whistle-blower access to decision-makers who could actively assist in resolution?

Condition 4: Last resort

The whistle-blower must be able to assess that any other action undertaken by them would be ineffective to resolve the situation of unethical and/or illegal activity uncovered by themselves.

Whilst these four conditions must be met if whistle-blowing is to be successful, I would respectfully suggest there is a *'neglected fifth'* condition of whistle-blowing that I have identified as 'personal commitment'. This fifth condition asks some very difficult questions of the potential whistle-blower that in turn require the whistle-blower to undertake intense personal reflection. They need to ask themselves the following questions before deciding upon their action:

- Are you willing to suffer financial hardship?

- Are you prepared to suffer professional hardship?

- Do you care that much for your community/professional body ?

- Are you prepared to pay the ultimate price for your whistle-blowing, are you prepared to lay down your life and/or that of your loved ones?

If a potential whistle-blower answers yes to all of the above, then they are well on the way to blowing the whistle, fully aware of the possible accompanying negative consequences of their actions and doing what they believe to be *'the right thing'* in the cause of trying to uncover and resolve situations of unethical and/or illegal activity(ies) and behaviours.

THE REALITIES OF WHISTLE-BLOWING

As can be seen, the act of whistle-blowing is not for the faint-hearted. It should never be undertaken on a whim and to be successful there must be a substantive and meaningful case of wrongdoing to be made by the whistle-blower. We have seen what conditions need to be met if successful whistle-blowing is to be undertaken, but the realities of this very brave step, and what must never be forgotten by those seeking to positively address unethical and/or illegal activity(ies) and behaviours of which they are aware, is that it is often accompanied by extreme personal and professional hardship and danger.

Whistle-blowing takes significant personal introspection before the act is undertaken and the whistle-blower must take time to understand their personal motives for doing so. Whistle-blowing must not be used as an act of **revenge** by the whistle-blower because if so, it is more than likely to be unsuccessful as the personal bias of the whistle-blower, at the expense of hard facts, will be all too evident to those tasked to seek a resolution. Whistle-blowing should also not be undertaken purely for **financial gain** as the credibility of the whistle-blower will be called into question, as will the merits of the case. In such instances the act of whistle-blowing stands the risk of being regarded not as the courageous act of somebody who wants to address a situation of unethical and/or illegal activity(ies) and behaviours but rather the act of a person seeking self-enrichment. Lastly, whistle-blowing is often the only means available to **protect the public and/or corporate interest** and whistle-blowers must genuinely seek to further these interests when faced with a situation of unethical and/or illegal activity(ies) and behaviours that does not comply with expected and often legally required activities and standards, nor does it uphold human dignity and rights.

Harris, Pritchard & Rabins (2009) propose that engineers have three responsibilities to the public when faced with a moral dilemma arising from unethical and/or illegal activities and/or behaviour arising in the field of science and technology. They build upon a concept put forward by Vannevar Bush (1980), who speaks of a contemporary conflict between science and democracy which he calls the *'democratic deliberation'*. As engineers work increasingly in a social context, it is suggested they have the responsibility to alert, inform and advise the general public and colleagues in other disciplines, as a means of ensuring that any type of whistle-blowing is well founded and supported and thus has every chance of success. These three responsibilities are detailed below.

Alert

Engineers have the responsibility to alert the public to the possible dangers arising from any unethical and/or illegal use of science and technology.

Inform

Engineers have the responsibility to inform the public of both the advantages and disadvantages of 'new' science and technologies to ensure that through insight, they can make an informed moral decision.

Advise

Engineers have the responsibility to advise and guide on issues concerning science and technology, especially to ensure understanding when there is consensus on an issue within the profession.

FAMOUS WHISTLE-BLOWERS

In recent times there have been many famous international whistle-blowers who have drawn attention to unethical and/or illegal activities and/or practices. Names such as Karen Silkwood, Linda Tripp, Frank Serpico, Jeffrey Wigands, Julian Assange, Bradley (Chelsea) Manning, Deep Throat (Mark Felt) and Edward Snowden immediately spring to mind. Their whistle-blowing activities have alerted the world to the illegal and/or unethical activities and behaviours of governments, organisations, individuals and industries.

However, whistle-blowing is not confined to America and Europe; we also have our own well-known cases of whistle-blowing in South Africa. A lack of provision of health services in Pollsmoor prison was highlighted by the whistle-blower Dr Paul Theron in 2007 and the industry collusion of price-fixing in the bread industry was brought to light by the whistle-blower Imraahn Ismail-Mukaddam and the Congress of South Africa Trade Unions (COSATU) in 2006.

The 'neglected fifth' condition of whistle-blowing in which personal commitment is required, has been all too evident in South Africa in recent times. Forfeiting one's life and/or that of family members in the pursuit of uncovering and bringing to light illegal and or unethical behaviour and practices has been an appalling consequence for those brave enough to 'blow the whistle.' Such was the case for Moss Phakoe **(https://www.dailymaverick.co.za/ opinionista/2018-08-27-the-price-of-speaking-out-as-a-whistle-blower-in-south-africa-is-high-just-ask-moss-phakoes-family/)** in March 2009, and Babita Deokaran **(https://www.dailymaverick.co.za/article/2021-12-10-six-accused-of-murdering-whistle-blower-babita-deokaran-in-court-for-bail-hearing/)** in August 2021, when their efforts to reveal unethical and/ or illegal activity directly resulted in their assassination by third parties. Each was gunned down in an attempt to prevent the 'truth' from being exposed.

Whistle-blowing was also a feature in revealing the facts in South Africa behind cases such as Life Esidimeni in 2015, and that of Jacques Pauw, the investigative journalist who blew the whistle on the Rwandan genocide and wrote the book *The President's Keepers* (2017) that detailed the part played by South Africa's State Security Agency (SASSA) in keeping the corrupt government of Jacob Zuma in power.

In line with the scope and context of whistle-blowing, there have been engineers who have blown the whistle on what they consider to be unethical and/or illegal activities and behaviours. Some of the seminal case studies that come to mind took place in America, such as the Bay Area Rapid Transport (BART) case study in 1969, Goodearl and Aldred versus Hughes Aircraft Corporation in 1990 and Roger Boisjoly, who warned of the O-ring problems that eventually led to the Challenger Disaster in 1986.

WHISTLE-BLOWING LAWS IN SOUTH AFRICA

South Africa passed the Protected Disclosures Act 26 of 2000, which attempted to codify the practice of whistle-blowing. It is commonly called 'the Whistle-blowers Act' and draws heavily on the UK's Public Interest Disclosure Act of 1998. The Protected Disclosures Act of South Africa (**https://static.pmg.org.za/docs/2006/060523guidelines1.pdf**) provides a useful guideline for employers with regard to the act of whistle-blowing, stating that it is a;

- **procedure** in terms of which any employee may disclose information relating to an offence or a malpractice in the workplace by his or her employer or fellow employee; and

- **protection** for an employee who has made disclosure in accordance with the procedures provided for by the Act, against any reprisals as a result of such disclosure.

The Act also provides that;

no provision in a contract of employment or other agreement which applies to an employer and employee may attempt to exclude any provision of the Act or –

- attempt to prevent an employee; or

- discourage an employee

from making a protected disclosure. Such provision (in a contract of employment) or agreement (between an employer and employee) has no legal effect.

There are also organisations in South Africa that provide both advice and support for those seeking to 'blow the whistle', having uncovered situations of either illegal or unethical activity and/or behaviour. Organisations such as the Ethics Institute of South Africa, The National Anti-Corruption Forum (NACF) and the Organisation Undoing Tax Abuse (OUTA) have all pledged their services in this very important area and the website **https://www. whistle-blowing.co.za** also offers a facility and advice for those wanting to blow the whistle. However, the South African politician Athol Williams, who was also a whistle-blower in the recent case of State capture, fled the country in 2021 after receiving threats to his life (**https://www.enca.com/news/athol-williams-raises-concerns-over-lack-protection-whistle-blowers**) and reminds us of the distinct lack of protection in South Africa for whistle-blowers despite Acts of Parliament.

HOW DO ORGANISATIONS AND PROFESSIONS MINIMISE WHISTLE-BLOWING?

The most effective way in which to reduce the chance of whistle-blowing in an organisation or a profession is to instil and practise a culture of corporate ethics. This can be achieved through the creation and application of the twin pillars of governance and compliance. If practised correctly, employees and members of professional institutions can be in no doubt of what is considered acceptable behaviour and/or activity. Governance and compliance are also assisted through clear and open lines of communication in the organisation or professional body in which the process of reporting unethical and/or illegal activities and behaviours is treated both confidentially (whilst under investigation) and with urgency. The professional and corporate codes put in place clearly state what is considered acceptable behaviour/practice, whilst also clearly stating the sanctions that accompany a breach of these either individually or collectively.

References

Bush, V (1980) *Science. The Endless Frontier: A report to the President on a Program for Postwar Scientific Research.* Washington: National Science Foundation.

Fleddermann, CB (2012) *Engineering Ethics.* Fourth Edition. US: Pearson Education, pp 108–110.

Harris, CE, Pritchard, MS & Rabins, MJ (2009) *Engineering Ethics. Concepts and Cases.* Fourth Edition. Canada: Wadsworth Cengage Learning.

Pauw, J (2017) *The President's Keepers.* Cape Town: NBPublishers.

LEARNING OUTCOMES

- What are the 5 conditions that must be met if whistle-blowing is to be successful?

- What are the 2 forms of whistle-blowing?

- What are the 2 types of whistle-blower?

- Name 3 conditions under which whistle-blowing should NOT be undertaken.

- List 3 famous whistle-blowers.

3
Engineering
Case Studies

EXTRINSIC FORCES AND THEIR DIRECT IMPACT ON ENGINEERING CASE STUDIES

13

Engineering case studies provide us with practical insights into the nature of engineering disasters. Their catastrophic results are invariably best understood and interpreted through using the ethical approaches we have examined earlier in this book in addition to considering the role and impact that both intrinsic and extrinsic forces can bring to bear on engineering outcomes. Whilst extrinsic forces are frequently a major influence or reason behind the engineering disaster and often cited in the case study forensics, they are invariably compounded by an accompanying lack or even absence of engineering ethics and a fatal absence of accurate and suitable risk and safety mitigation strategies. Finally, a lack of engineering professionalism has been cited in some engineering disasters, when the duties and virtues expected to be demonstrated and practised in engineering activities and engineering project management have been lacking or, in some cases, have been totally absent.

Table 13.1 provides a limited list of some well-known international engineering case studies and the principal extrinsic forces that can be directly associated with their unfortunate engineering outcomes. (There can be other extrinsic forces such as politics and inadequate law(s) that can also be seen to directly impact engineering outcomes including some of the case studies listed below.) Interestingly, in many cases of engineering disaster, however, it is a combination of extrinsic forces, lack of ethics, lack of engineering professionalism and absence of suitable risk and safety mitigation strategies that serve to create the tragedy.

Table 13.1: Extrinsic forces associated with engineering disasters.

EXTRINSIC FORCE	ENGINEERING DISASTER
WEATHER	**Space Shuttle Challenger Disaster:** The 'O' rings malfunctioned due to cold temperatures. **The Boston Molasses Disaster:** The storage tank for molasses collapsed due to hot weather.
ECONOMICS	**The Ford Pinto Disaster:** Strict application of a cost–benefit analysis.
HUMAN ERROR	**The Chernobyl Disaster:** Due in part as a result of human error in testing No 4 Reactor.
EQUIPMENT FAILURE	**The Bhopal Disaster:** Due in part to a lack of equipment maintenance in the plant that compromised safety. **Apollo 1 Disaster:** Due in part to the hatch cover of the cabin not being quickly removable under high pressure.
CORPORATE STRUCTURE	**Hyatt Regency Walkway Collapse:** Inadequate corporate structure without clear lines of accountability and responsibility for designs and construction by subcontractors. **Boeing 737 MAX 8:** The aircraft crashes were attributed to the MCAS self-activating. Boeing's relationship with the Federal Aviation Authority (FAA) allowed them to produce self-certified data that in turn assisted the FAA with issuing airworthiness certification. The Boeing corporate structure relied on reputation, deadlines and sales to ensure corporate profits.

The following case studies of engineering disasters are in the main drawn from South Africa and serve as examples of the importance that should be placed on ethics, engineering professionalism, the existence of extrinsic and intrinsic forces and the need for risk and safety mitigation strategies by engineers when undertaking engineering activities, engineering projects and engineering designs.

CASE STUDIES

KARIBA DAM COLLAPSE AND MERRISPRUIT DAM DISASTER

14

The following case studies, Kariba Dam Collapse and Merriespruit Dam Collapse, demonstrate how an extrinsic force – weather – has in these two engineering projects directly impacted engineering outcomes. Case studies like this also serve to demonstrate the importance of appropriate safety and risk mitigation surveys and strategies that need to be undertaken by engineers if disaster is to be averted. Moreover, of interest is that when details of disasters such as these and others of an environmental nature are examined, an absence of engineering ethics is often uncovered in addition to an absence of an understanding of human dignity in the associated engineering activities and practices.

 CASE STUDY: THE KARIBA DAM COLLAPSE

Lake Kariba (**https://www.britannica.com/topic/Kariba-Dam**) is the largest artificial (manmade) lake and reservoir by volume in the world. It was built on the border of what is now known as Zambia and Zimbabwe (formerly Northern and Southern Rhodesia in colonial times). It has a surface area of 5 580 km^2 and a water storage capacity of 185 km^3. It has an average depth of 29 m and at its deepest is 97 m. It is over 223 km long and up to 40 km at its widest point. It was constructed to dam the great Zambezi River floodplain.

The design of the Kariba Dam was undertaken by Andre Coyne, a specialist engineer and the inventor of the *'arch dam'* design. Kariba Dam was designed and built with what is known as a *'double curvature concrete arch dam'* that stands 128 m (420 ft) tall and 579 m (1 900 ft) long. An Italian company Impresit, familiar with such engineering projects, was awarded the construction contract.

At the time of its commission the Kariba Dam project was the largest hydro-electric project in Africa and it was sited where the Zambezi River flows through the natural Kariba Gorge in the lands traditionally inhabited by the Tonga tribe. The picture below shows the Kariba Dam wall under construction.

Source: https://ejatlas.org/conflict/kariba-dam-zambia-zimbabwe

ENVIRONMENTAL CONCERNS

The chosen site of the Kariba Dam was home to the Tonga tribe, who underwent forced removal and became what has been described as *'development refugees'*. The irony of their forced removal, in light of the benefits the Kariba Dam was purported to be able to provide for the citizens and industry of both Zambia and Zimbabwe, has been chillingly described by the *New York Times* (**https://www.nytimes.com/interactive/2020/07/22/magazine/zambia-kariba-dam.html**) as follows:

[T]he building of the Kariba Dam redirected enormous wealth to colonial parties at the expense of the rightful dwellers of the Gwembe Valley, who are now considered 'development refugees' and lack adequate access to water and electricity.

A total of 57 000 Tonga (**https://www.herald.co.zw/the-tale-of-kariba-dam-nyaminyami/; https://www.zambiatourism.com/destinations/lakes/lake-kariba/history/**) tribespeople distributed across 193 villages were the human subjects of forced resettlement for the Kariba Dam project, also forced to relinquish their ancient way of life which was agrarian-based and relied on the seasonal floods provided by the Zambezi for their crops. In addition, their sacred ancestral burial sites and family homes were to be submerged under water, never to be seen again. Interestingly the tribal elders foretold only doom, should the flow of the river be stopped by the construction of a dam wall. Their indigenous knowledge and ancient legends and mythology led them to believe the god of the Zambezi River Nyaminyami, a water snake, would retaliate should the Zambesi River be blocked and its anger would result in either pushing the dam wall over, or alternatively it would bypass the wall and creep around the sides of the dam wall construction and cause floods.

However, it was not just tribespeople that needed to be relocated for the construction of the Kariba Dam wall; wildlife also had to be removed from the area and relocated to Matusadona National Park to make way for the largest manmade lake in the world. Over the course of the five years it took to construct the Kariba Dam, conservationists engaged in a relocation project named *'Operation Noah'* that managed to capture and relocate at least 6 000 wild animals including rhinoceros, hippopotamus, baboons, elephants, lions, buffalo, leopards, zebras, warthogs and antelopes.

Construction began on the dam in 1955 and was completed in 1959 on government *'federal colony'* orders at a reported cost of US $480 million. The picture below shows the completed dam wall.

Source: http://www.zimbabweconnections.com/lake-kariba/

During the planning and designing of the Kariba Dam, engineers consulted weather records and noted existing weather patterns to establish the flow of the Zambezi River. Records showed the Zambezi River experienced maximum flow in March/April with an annual average flow of 7 000 cubic metres per second. The odds of successive floods were calculated at around 1 000 to 1. However, this chance occurrence happened during the construction of the dam.

The first flood occurred in late 1956. The water rose up to 20 m and flooded the cofferdam (a structure used to create a watertight fence permitting the construction of piers and other hydraulic work associated with modern dam construction). Once the flood water had subsided, engineers took the decision to build a higher second cofferdam to ensure that it could contain any increased water levels. However, it proved insufficiently high when in 1958 another unexpected flood took place (which remains the highest on record for the area), in which the flow of the Zambezi reached close to 16 000 cubic meters per second. This translates into an increase in river flow of over 800% higher than the average and was calculated at 16 million litres per second. This flood, described as being of *'biblical proportions'*, was calculated to occur only once every 10 000 years. It also created a waterfall in excess of 8 m that cascaded over the second, higher cofferdam and swept scaffolding away and damaged some of the recently constructed infrastructure of the dam wall. Undeterred, the construction company proceeded to clear up the damage from the floods and relentlessly continued construction to meet project deadlines. In a further freak accident, 17 construction workers (11 of whom were Italian) fell into a hole and were buried in wet concrete. It was decided the cost of reclaiming the dead workers was too high and a decision was made by the chief engineer to leave their bodies within the dam wall. This further added to the local belief of the project being *'cursed'* and was the consequence of the river god Nyaminyam's rage because of the natural flow of the Zambezi River being dammed.

The picture below taken just before the 1958 floods illustrates how exposed to danger the construction workers were on this project. It is reported that a total of 90 construction workers died in the building of the Kariba Dam wall.

Source: https://af.wikipedia.org/wiki/L%C3%AAer:Kariba_Dam_Construction_2.jpg

LEARNING OUTCOMES

- Explain how human dignity was denied in this construction project.

- This project was approved on a cost–benefit basis. What costs were not calculated?

- Were adequate safety and risk mitigations undertaken for this project? Explain your answer.

- Were the human rights of the local tribes recognised? Explain your answer.

- Explain which other extrinsic forces aside from weather could be argued to directly impact the outcomes of this engineering activity?

CASE STUDY: THE MERRIESPRUIT DAM DISASTER

In 1978 Harmony Gold Mine designed and built Merriespruit Dam No 4 at Harmony Gold Mine in Virginia in the Free State in South Africa (**https://www.tailings.info/casestudies/merriespruit.htm**). It was a *'tailings dam'* and was built within 300 m of the first house in a residential suburb called Merriespruit. The suburb consisted of 250 houses that were built a considerable time before the dam was constructed and were family homes.

DETAILS OF THE DAM BUILD AND ITS USE

The Merriespruit Dam was built using the semi-dry paddock method, which was an accepted method of construction for a dam that was built to act as a depository and processing facility for gold tailings. In chapter two of her dissertation, Wortman (**https://repository.up.ac.za/bitstream/handle/2263/23079/02chapter2.pdf?sequence=3&isAllowed=yp2–14**) states that:

> An average gold mine is estimated to produce roughly 100 000 tons of tailings per month over an expected lifetime of twenty-five years.

The tailings were transported by heavy goods vehicles to the Merriespruit Dam during the day to settle. They were then processed in the dam at night. In the middle of the dam there was a drain (penstock) to get rid of all excess water from the processing of the gold tailings, including excess natural rainfall.

Figure 14.1 provides a graphic for the semi-dry paddock method and illustrates the water gain and loss in a typical tailings dam.

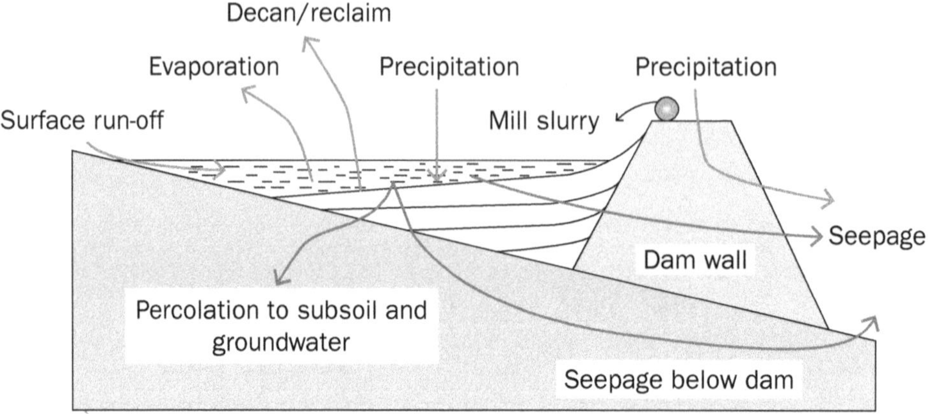

Figure 14.1: The semi-dry paddock method.
Source: https://www.sciencedirect.com/topics/engineering/tailing-dam

However, on 22 February 1994 the Merriespruit Dam No 4 collapsed as a consequence of a thunderstorm in which 55 mm of rain fell in 30 minutes **(https://mg.co.za/article/1996-03-22-scars-of-the-survivors/)**. The 31 m high northern dam wall was breached, causing a 2.5 m wave of water, sediment and slurry (an estimated volume of 600 000 cubic meters) from the gold tailings processing. This *'wave'* travelled nearly 4 km and covered 2 km² of the residential suburb of Merriespruit in its path. Eighty homes in the suburb were destroyed beyond repair, whilst 170 homes were severely damaged. A total of 17 lives were lost in the disaster.

PROBLEMS LEADING UP TO THE DISASTER

A year before the disaster occurred (1993), a leak in the Merriespruit No 4 dam was reported. After this, all deposits of tailings were halted in that compartment of the Merriespruit Dam No 4. The dam was reportedly in an unacceptable condition at the time of the disaster and did not have sufficient and mandatory freeboard to retain the rainfall from a once-in-a-hundred-years 24-hour thunderstorm. Moreover, water flowing over the top of the dam wall causing streams of water (known as flurry water) to flow into the suburb of Merriespruit had been reported by residents before the disaster occurred and upon inspection by contractors and Harmony

Gold employees, the Merriespruit Dam was insufficiently repaired due to costs and also because skilled employees in the maintenance of paddock dams had been replaced with personnel with fewer skills, less knowledge and far less experience.

THE OFFICIAL FINDINGS AFTER THE DISASTER

An official investigation was undertaken involving both the Minister of Justice and the State. The investigation discovered that despite legislative requirements being in force for this type of construction, the Merriespruit Dam No 4 was not maintained to the legislative standards. Considering this, Harmony Gold Mine was found guilty of culpable homicide and its contractors were found to be negligent in both the maintenance of Merriespruit Dam No 4 and its operation. In an effort to ensure justice and to prevent a disaster of this nature happening again in South Africa, Goldfields, the contractors for the maintenance and operation of Merriespruit Dam No 4 and six employees were all given fines.

The Claasens family, who were living in Merriespruit at the time of the disaster, lost their 14-year-old daughter and their younger son was left with severe, life-changing injuries after their home was buried in the wave of *'sludge'* that flooded their home and the suburb of Merriespruit (**https:// mg.co.za/article/1996-03-22-scars-of-the-survivors/**). They received R10 000 in compensation from Harmony Gold for their loss and had to personally fund all medical bills for their son and themselves resulting from the disaster. They and other survivors of the disaster took a decision to pursue a civil court action against Goldfields in an attempt to receive compensation for loss, personal injury and earnings.

RESOLUTIONS TAKEN

Lessons were learned after the Merriespruit Dam disaster and changes were made in the laws governing safety with a new code of practice for mine residue deposits to ensure that another disaster of this nature could never happen again in South Africa. The Council for Scientific and Industrial Research (CSIR) was instrumental in undertaking the research required to be able to draft a law with suitable standards. It became known as the Draft Code of Practice for the Design, Operation and Closure of Tailings Dams.

As a result, today all active tailings dams are fitted with ring mains that carry excess water to a return-water dam that can manage sudden or excessive water flows, whatever their reason, from penstocks. In conjunction with this, mandatory freeboards have been increased to ensure the perimeter walls cannot be breached/overwhelmed.

In addition, the law now mandates the maintenance of such dams cannot be delegated to contractors and/or individuals with inadequate experience and their construction by accredited builders must now have an approved *'hazard management strategy'*. This essentially forms an integral part of a safety and risk assessment with an agreed mitigation strategy for disaster management.

Finally, legislation now requires that no dam can be built within 1 km of a residential area, nor can it be built higher than nearby residences.

LEARNING OUTCOMES

- Explain how Harmony Gold denied the residents of Merriespruit their human dignity and their human rights.

- Discuss how the categorical imperative of an engineer was not met by those working at the Merriespruit Dam No 4.

- Safety and risk mitigation strategies in addition to compliance with accepted standards and practices are critical in engineering actives. Discuss how their absence contributed to the Merriespruit Dam No 4 disaster.

- List the extrinsic forces you feel further contributed to this engineering disaster.

- Were environmental ethics considered by Harmony Gold during the siting and maintenance of Merriespruit Dam No 4 and the processing of the gold tailings?

SHELL EXPLORATION OFF THE WILD COAST AND SOUTH DURBAN BASIN

CHAPTER

15

The following two case studies can be seen to be examples of a lack of environmental ethics by international organisations operating in South Africa. In both case studies we are made aware of the direct impact engineering activities can have not only on the environment but also on the surrounding communities and residents.

 CASE STUDY: SHELL EXPLORATION OFF THE WILD COAST

To deepen its strategic thinking and accurately predict future energy trends, Shell uses scenario planning (**http://www.shell.com/scenarios**) that considers the direct impact that economics, politics, social change and science have on energy types and consumption. In so doing, the company often looks beyond the traditional energy outlooks and considers alternative points of view to understand the global energy system and environment. They are committed to fostering policies that;

promote the development and use of cleaner energy and improve energy efficiency.

Shell anticipates the work they have undertaken in terms of their strategic thinking will provide them with corporate direction. In addition, it is hoped their research will provide the information for their executive decision-making for at least the next 40 years, as they seek to narrow the gap between energy supply and demand. In so doing, Shell is ever mindful of the environmental concerns that unveil themselves over time through the progress of science and technology.

What has become apparent is that an increasing global population and its simultaneous urbanisation and industrial lifestyle have resulted in a significantly increased demand for energy not just in terms of personal use, but for commercial growth too. Shell predicts that by the year 2050, total global energy consumption will be some 80% higher than it is today. Considering this prediction, Shell proposes that energy strategies will need to recognise and accommodate the challenges this will present.

Shell, in undertaking its corporate strategic thinking, developed two 'vistas' which it believes present likely future scenarios (**http://www.shell.com/scenarios**) in the coming years which it has named 'Mountains' and 'Oceans'. In the former scenario, Mountains, it cites the importance of 'new shale and tight gas' resources that will be able to meet energy demands and decrease the developed world's existing dependence on oil in addition to adopting alternative energy sources such as solar. This can be achieved by Shell due to environmental vision and supply-side incentives. In the latter scenario, Oceans, Shell recognises the environmental impact of fossil fuels, such as coal, in developing economies, but notes this provides grounds for a significant lag between adopting 'new' cleaner energy owing to excessive costs passed on to the energy consumers in the form of a carbon-neutral capability.

Of interest is that Shell has stated its role in providing energy both now and in the future is;

most importantly about helping people take a journey that guides them into better choices based on richer considerations of the world around them.

SHELL'S EXPLORATION OFF THE WILD COAST

In view of their energy strategy and scenario planning, exploration to establish suitable 'new shale and tight gas' supplies are critical to the future of Shell as a profit-generating energy corporation. Seismic surveys are the means chosen and used by Shell to establish the presence and quantity of 'new shale and tight gas' for future extraction.

In 2014 Impact Africa (a company with which Shell was working to undertake the seismic survey) was granted an exploration right by Minister Gwede Mantashe of the Department of Mineral Resources and Energy (DMRE), who held talks with traditional leaders to elicit their consent. The exploration right for Shell was renewed by the Minister in 2017.

However, lobbyists and activists shared their concerns that *'critical considerations'* were ignored by government before the granting of exploration rights to Shell, neither was adequate public engagement undertaken. Furthermore, News24 **(https://www.news24.com/fin24/ companies/wild-coast-communities-want-court-to-make-final-call-to-block-shells-seismic-survey-20220503)** reported that neither adequate or sufficient public engagement was undertaken in the Wild Coast communities by Shell or Impact Africa and suggested that if public forums had been held, expert witness statements from both Shell and the community could have been made. This would have ensured that concerns with regard to the environmental impact of the proposed seismic surveys were both discussed and addressed by all parties such as the potential harm that seismic activity could have on the biodiversity of the area as well as the accompanying negative impact it could have on the livelihoods of local and indigenous residents of the Wild Coast.

FACTS INCLUDED IN THE HIGH COURT ACTION

News 24 **(https://www.news24.com/fin24/companies/wild-coast-commu-nities-want-court-to-make-final-call-to-block-shells-seismic-survey-20220503)** reported that civil society organisations such as Greenpeace, Sustaining the Wild Coast, Wild Coast Residents and six other applicants filed their heads of argument with the Eastern Cape Division of the High Court in an attempt to be granted an interim interdict to stop Shell's planned seismic survey off the Wild Coast. They maintained such activity would cause *'irreparable harm'* to marine life along this immediate coastline and beyond. Moreover, these applicants further contested the blasting would negatively;

> impact the livelihoods and the constitutional and customary rights of coastal communities.

They further argued that when the original rights to exploration were initially granted in 2014 to Impact Africa, the required environmental authorisation according to the National Environmental Management Act (NEMA) was not obtained. The applicants also contested that public engagement was not correctly undertaken. They implied that full disclosure was not provided to local communities who were inadequately *'consulted'* not only in terms of the possible damage the seismic activity could have upon their livelihoods, but also the possible damage it could have on their spiritual and cultural practices too.

Furthermore, information concerning Impact Africa's proposed seismic activity was distributed electronically – a medium that local communities did not have access to – using either English and/or Afrikaans as the language of communication, neither of which many in the local communities read or spoke. Most Wild Coast residents speak isiXhosa or isiMpondo and neither language was used to communicate the proposed seismic activity off the Wild Coast. Even more alarming, it was claimed that Impact Africa chose the medium of newspapers to communicate their proposed seismic coastal activities rather than local radio. The latter is the preferred communication medium by local indigenous residents and one which could easily have been used as it not only has the benefits of language and immediacy but also could have been used as a critical means of communication for the provision of a *'public information service'* with regard to the proposed seismic engineering activity. Furthermore, if local radio had been used, it could have served as a forum for public engagement in which expert witnesses, the community and Shell could have engaged with each other and shared facts and concerns surrounding the planned seismic activity for what is considered an *'ecologically sensitive'* area.

In August 2021 the DMRE once again renewed the exploration right for Shell to undertake seismic surveys off the Wild Coast.

HIGH COURT VERDICT

On 1 September 2022 it was reported in the *Daily Maverick* (**https://www. dailymaverick.co.za/article/2022-09-01-activists-celebrate-big-win-for- wild-coast-against-mantashe-and-shell/**) that the seismic study proposed by Shell off the Wild Coast was interdicted:

> The Makhanda High Court has set aside a decision made by the Department of Mineral Resources and Energy to grant an exploration right to Shell to conduct seismic surveys off the Wild Coast.

The 2014 exploration rights granted by the DMRE to Impact Africa and Shell were considered by the presiding Judges to be *'procedurally unfair'*.

Pooven Moodley, a director at Natural Justice (one of the applicants), is reported to have said:

> The court was clear that communities need to be properly consulted and that environmental impact assessments are critical. The cultural and spiritual connection to the land and the ocean featured strongly in the judgement.

In this instance *'environmental justice'* can be seen to have been achieved and civil society was able to claim a victory for local communities and the environmental sustainability and biodiversity of the Wild Coast.

LEARNING OUTCOMES

- Explain how Shell could be accused of not following the basic steps expected from engineers in terms of the practice of public engagement.

- Was there an absence of consideration of human dignity and human rights in this case study? Discuss.

- Were the environmental duties expected of engineers and outlined in the ECSA Code of Ethics met by Shell's engineers?

- Could Shell be accused of adhering to a strict cost–benefit strategy when seeking to undertake seismic surveys along the Wild Coast? Discuss.

- Did the proposed engineering activity (seismic survey) comply with the Utilitarian tradition? Discuss.

CASE STUDY: SOUTH DURBAN BASIN

Source: https://www.sabcnews.com/sabcnews/durban-south-basin-a-hotspot-for-high-air-pollution/

THE BACKGROUND TO THE SOUTH DURBAN BASIN COMMUNITY AND INDUSTRIES

The South Durban Basin (SDB) is the industrial hub of Durban. It is home to +/- 120 industries including petrochemical refineries (Engen and SAPREF being chief amongst them), chemical industries, paper manufacturing (Sappi and Mondi), car manufacturers and toxic landfill companies. In the 1950s the enforcement of the Group Areas Act in South Africa displaced black, coloured and Asian people to the SDB area where they experienced poor living conditions and environmental pollution. To this day it remains the home for low-income citizens of colour who continue to bear the costs of the petrochemical and chemical industries. The term often used for these situation and others like it around the world is *'environmental racism'* and in contemporary times is best understood as an *'illegal exaction'* upon the poor. Research undertaken by SRK Consulting (**https://www.accord.org.za/ ajcr-issues/environmental-conflicts-in-the-south-durban-basin/**) in 2004 identified 52% of the adult population of the SDB was not economically active, and 79% of those who were economically active earned less than R15 000 per annum.

The 1996 population census recorded that 250 000 people lived in the SDB in the residential areas of Merebank, Bluff, Clairwood, Isipingo, Umlazi, Lamontville and Wentworth. This figure had increased to 400 000 by 2002 (according to research undertaken by the Council for Scientific and Industrial Research (CSIR)) **(https://www.sabcnews.com/sabcnews/durban-south-basin-a-hotspot-for-high-air-pollution/)**. This resident population has continued to suffer from high air pollution levels over many years because it has the;

 largest concentration of petrochemical industries in the country.

Jaggernath **(https://www.ajol.info/index.php/ajcr/article/view/63316)** confirms the presence of;

 emissions of unacceptable levels of toxins, chemical waste and a large content of sulphur dioxide [are present and measurable in the SDB because of the] characteristics of the industrial processes and activities.

Residents in the SDB live extremely close to petrochemical and chemical industries and breathe the polluted air day and night, in addition to enduring the foul smell of sulphur dioxide **(https://www.spotlightnsp.co.za/2020/09/02/covid-19-we-cant-breathe-residents-say-in-south-durban-caught-between-covid-19-and-polluted-air/)**. Their close proximity to petrochemical and chemical industries and the pollution caused by them, not just by emissions but also poor quality of the drinking water and soil (as streams and canals are polluted by industries in the SDB), results in residents suffering from upper respiratory disease, chronic obstructive pulmonary disease, asthma, skin rashes, burning eyes, itchy throats, kidney and liver malfunctioning and on-going stomach upsets. There is even a suggestion of a direct correlation between the rates of cancer, leukaemia and high pollution levels. However, the on-going high levels of pollution in the SDB have been further exacerbated by numerous gas leaks from the petrochemical industries, the most recent being a methyl mercaptan gas leak at the Engen refinery in July 2020, which the company confirmed had taken place.

In addition to the high levels of pollution in the area, residents suffer the negative environmental consequences from being located on a flood plain with residential and commercial development in close proximity to the coastline. SDB residents, as well as being victims of environmental pollution, are now also victims of climate change impacts such as floods, extreme storms and excessive winds.

THE CONTINUED SEARCH FOR ENVIRONMENTAL JUSTICE

Environmental justice (**https://www.accord.org.za/ajcr-issues/environmental-conflicts-in-the-south-durban-basin**/) has been defined by South Africa's Environmental Protection Agency in 1998 as:

> The equal treatment and participation in decision-making of all people regardless of their race, colour, nationality or income status.

Furthermore, in an article published by the *Daily Maverick* (**https://www.dailymaverick.co.za/article/2019-06-28-environmental-injustice-in-south-durban-community-caught-between-toxic-polluters-and-climate-shocks**/) in which it identified the lack of environmental rights enjoyed by the residents of the SDB it quoted section 24 of the South African Constitution that states that all South African people irrespective of their race, colour or ethnic differences have the right to an environment that is not harmful to their health or well-being. It further advises the environment needs to be protected not only for the benefit of the current generation but also for future generations, through the use of appropriate legislation in the areas of sustainable development, conservation and justifiable social and economic development.

The Merebank Ratepayers Association have persistently undertaken public engagement with regard to tackling the high levels of pollution and emissions in the SDB, addressing the oil refineries in apartheid South Africa and in the democratic South Africa. They have issued memorandums of complaint to the refineries, who initially denied they were polluters and suggested the high emissions results were due to the residents living *'downwind'* of their industries and others active in the SDB area. A distinct lack of enforceable environmental laws in apartheid South Africa created a stalemate between the residents of the SDB and the refineries, not least because under apartheid law the refineries operated under the Official Secrets Act and were surrounded by razor wire and protected by guard towers. This arguably seemed to offer them an air of invincibility and protection which seemed to reflect in their corporate attitude towards the resident communities in the SDB.

However, in 1995 residents of the SDB created a Good Neighbour Agreement (GNA) which formalised their demands to industry. Whilst industry initially ignored this community initiative, it was followed up in 1997 by the SDB Environmental Alliance, an activist/lobbyist group which gained both support and traction. This was just before the National Environmental Management Act (NEMA) was promulgated in 1998, formalising the government's position in terms of environmental laws and standards and further endorsing the demands of community members in SDB seeking environmental justice.

Eventually, Engen agreed to reduce sulphur emissions and pollution in general by 65%, starting from 1999. However, the environmental degradation of the area and the pollution remains. Joint initiatives between the SDB community, government and industry are on-going.

LEARNING OUTCOMES

- The petrochemical industry operating in the SDB could stand accused of lacking moral virtue. Discuss which moral virtues these might be.

- The demands made by the residents of SDB were grounded in environmental intrinsics. Explain.

- Has human dignity been denied the residents of SDB by industries operating in their environs? Discuss.

- Explain why the engineering industry has a duty to maintain an environmental balance.

- Explain how human rights were denied and are arguably currently being denied to the residents of SDB.

CHAPTER 16

FORD SA KUGA AND FORD PINTO

The following case studies both concern the automaker Ford. Whilst the Ford Pinto case study did not occur in South Africa, it remains a seminal case study for engineering ethics for the negative consequences that can occur if a strict cost–benefit analysis is used. The Ford Kuga case study, on the other hand, did occur in South Africa and unfortunately is a contemporary example in which Ford South Africa have been accused of acting unethically.

CASE STUDY: FORD SA KUGA

The Ford Motor Company wanted to compete worldwide in the valuable sports utility vehicle (SUV) segment of the automotive market. As a means of achieving this, Ford designed the Ford Kuga as an affordable family-sized SUV complete with the latest Ford technology. The model range was to include a 1.5, 1.6 and 2.0 litre engine size. The new Ford models sold well globally, enabling the Ford Motor Company to compete effectively in the expanding SUV market and gain significant market share.

THE FORD KUGA IN SOUTH AFRICA

In 2015 Reshall Jimmy, aged 33, died when his Ford Kuga caught fire whilst he was on holiday in the Western Cape. He drove a 1.6 litre (1.6L) Ford Kuga with an Ecoboost engine and remains the only death in South Africa due to a fire in a Ford Kuga. However, in South Africa, there are more than 46 incidents on record in which the 1.6L Ford Kuga Ecoboost model caught fire.

The images below show Ford Kuga SUVs that have spontaneously caught alight in South Africa and of interest is the owners of the vehicles in every case had maintained their vehicles and were not driving recklessly or at high speed. Initially, Ford South Africa denied responsibility for the fires in the Ford Kuga models.

Source: https://www.bbc.com/news/world-africa-38641489

Source: https://www.news24.com/you/Archive/ford-sa-recalls-its-16-litre-kuga-model-20170728

However, whilst Jeff Nemeth, CEO of Ford South Africa (Ford SA) at the time was quoted by News24 **(https://www.news24.com/you/Archive/ford-sa-recalls-its-16-litre-kuga-model-20170728)** as saying;

at Ford, the safety of our customers is our number one priority,

events and experiences with the Ford Kuga 1.6L Ecoboost model in South Africa initially demonstrated quite the opposite, until a wave of public opinion caused by reports of Ford Kugas catching alight caused the company to review the situation.

Ford SA spokesman Rella Benardes was reported in TimesLive **(https://www.timeslive.co.za/news/consumer-live/2016-12-22-ford-confirms-kuga-fires-confined-to-single-model-concedes-engine-overheating-a-possible-cause/)** to have assured owners of affected Ford Kugas that, as part of a mitigation strategy:

Dealers will check the coolant concentration level and for any leaks or damage to the cooling system, plus conduct cooling system pressure tests.

It was felt by Ford SA this effort should serve to allay any fears Ford Kuga owners might have about their vehicles catching alight.

FORD SA INVESTIGATION AND THE FAMILY INVESTIGATION INTO RESHALL JIMMY'S DEATH

The National Consumer Commission (NCC) of South Africa imposed a deadline on Ford SA of 15 months from the date of Reshall Jimmy's death in the blaze which engulfed his Ford Kuga1.6L EcoBoost to establish the cause of that blaze and establish if there were any associated links it might have with other fires reported in similar Ford Kuga models in South Africa. The NCC also suggested that Ford SA should advise Kuga owners in South Africa of precautionary steps they would take if the fires were found to be due to a mechanical/design error that could be seen to represent an absence of care and/or responsibility on behalf of Ford SA.

A forensic investigation into Reshall Jimmy's death in his 2014 Ford Kuga 1.6L Ecoboost model was undertaken both by Ford SA and the family of the deceased. *TimesLive* (**https://www.timeslive.co.za/news/consumer-live/2016-12-22-ford-confirms-kuga-fires-confined-to-single-model-concedes-engine-overheating-a-possible-cause/**) confirmed the only way the family were able to identify the body that was trapped inside the burnt vehicle was through a DNA sample taken from Jimmy's brother for comparison. After their forensic investigation Ford SA maintained the fire in Jimmy's Ford Kuga started in the rear of the vehicle, despite the evidence of two independent witnesses who stated in their affidavits, the fire began at the front of the vehicle and a video shot by a passer-by confirmed their version of events. The family maintained in subsequent court hearings the evidence they had gained from their investigation repudiated the automaker, who argued the engineering employed in the Ford Kuga was not in any way responsible for the death of Reshall Jimmy in his Ford Kuga blaze.

David Klatzow was the forensic pathologist/investigator hired by the Jimmy family and his findings were at odds with the findings of Ford SA. He found the fire had started at the front of the car and was also able to prove that Ford SA's claims that Reshall Jimmy had been shot in the head and murdered were false, as was their claim that he had committed suicide. Klatzow found no external injuries to the victim. The post-mortem report showed no evidence of gunshot wounds and further established the hole in the head of the victim was the result of the skull cracking in the heat of the vehicle blaze. He also attested the post-mortem report found evidence of soot in Reshall Jimmy's trachea, suggesting the victim was still alive during the fire, which would directly dispute the notion of suicide put forward by Ford SA.

In May 2017 (**https://www.engineeringnews.co.za/article/fords-r1m-offer-a-slap-in-the-face-says-reshall-jimmys-family-2017-05-25/rep_id:4136; https://ewn.co.za/2017/05/25/family-of-deceased-ford-kuga-victim-wants-fair-compensation**) it was reported by the Jimmy family that Ford SA had made them an offer of a new Ford car worth R1 million if they dropped their civil suit against Ford SA and did not pursue the case any further. The family refused the offer from Ford SA, finding it both insulting and inappropriate. It has been reported that Ford SA have still not admitted responsibility for the fire that killed Reshall Jimmy in his Ford Kuga.

FORD'S RESPONSE TO THE KUGA FIRES

It was reported by *TimesLive* (**https://www.timeslive.co.za/news/consumer-live/2016-12-22-ford-confirms-kuga-fires-confined-to-single-model-concedes-engine-overheating-a-possible-cause/**) there were an estimated 6 300 Ford Kuga owners in South Africa whose cars were bought between December 2012 and October 2014 and who could be affected by the issue of fires in their Ford Kuga vehicles. However, Ford SA were accused of dragging their heels in trying to identify the issues affecting the Ford Kuga 1.6L Ecoboost model that might provide clarity on the cause of the fires. In January 2017 News24 (**https://www.news24.com/you/Archive/ford-sa-recalls-its-16-litre-kuga-model-20170728**) reported that Jeff Nemeth had eventually conceded the following:

> Based on current data, the fires are due to over-heating caused by a lack of cooling circulation which can lead to a cracking in the cylinder head and therefore an oil leak. If the leaking oil reaches a hot engine component, it can potentially catch fire.

FORD KUGA RECALL AND COMPENSATION

In 2017 this led to Ford SA issuing a company statement (**https://www.ford.co.za/about-ford/newsroom/2017/ford-issues-safety-recall-for-kuga-1-6/**) undertaking a major safety recall of 4 556 vehicles that would be carried out by their Ford dealerships across South Africa. The company told consumers the vehicle check would take approximately one hour. Ford SA also took a decision to expand their roadside assistance in South Africa with the Automobile Association (AA) for all Ford Kuga 1.6L models and simultaneously extended the maintenance cover on all Kuga models sold in South Africa to a total of six years or 200 000 km, whichever came first, at no cost to the consumer.

However, it was felt by Kuga owners in South Africa that Ford SA had not reacted quickly enough to both the evidence and their concerns about the safety of their vehicles and neither had they recognised the lack of financial equity they had in their cars, which, owing to the fire hazard they were seen to present, would not retain value in the South African used car market.

The NCC (**https://www.gov.za/speeches/national-consumer-commission-slaps-ford-motor-company-southern-africa-fine-r35m-29-nov-2019#:~:text=The%20National%20Consumer%20Commission%20('NCC,of%20the%20Consumer%20Protection%20Act**) eventually entered into a settlement agreement with Ford SA and the details were shared in Parliament. In 2020 it was reported by IOL (**https://www.iol.co.za/news/south-africa/gauteng/ford-compensates-sa-owners-for-kuga-suv-fires-fe6df44c-4f70-40db-85a3-d50e56d5e205**) that Ford SA had now agreed to compensate 47 owners of Ford Kugas that were damaged by fire. If compensation was accepted, a payment of R50 000 would be paid to each vehicle owner in full and final settlement.

LEARNING OUTCOMES

- Did Ford SA put their brand reputation before their engineering duty? Discuss.

- Explain how Ford SA could be accused of neglecting human dignity and human rights.

- Discuss whether Ford SA demonstrated engineering professionalism in their management of the safety and risk issues surrounding the Ford Kuga 1.6L Ecoboost.

- List which moral virtues Ford SA failed to apply in their response to the dangers associated with the Ford Kuga.

- Could it be said the lack of an appropriate response by the Ford Motor Company to the safety and design issues surrounding the Kuga 1.6L model was a function of the problem being experienced in Africa rather than Europe or America?

CASE STUDY: THE FORD PINTO

In the late 1960s the Ford Motor Company US division needed to expand their market share in the subcompact car market with a view to dominating it with the launch of new models. Ford US decided the Pinto model was going to spearhead their efforts for market dominance. Design for the Pinto started in late 1968, with the Ford president Lee Iacocca rushing the project launch to meet an early 1971 deadline (**https://philosophia.uncg.edu/phi361-matteson/module-1-why-does-business-need-ethics/case-the-ford-pinto/**). The usual time to design, test and launch a Ford model was four years. The Ford Pinto was designed, manufactured and launched in three. The Ford Pinto became the largest selling subcompact car in America.

FORD PINTO'S PRE-PRODUCTION TESTING

The designers of the Ford Pinto model had certain critical restrictions placed upon them by Ford management. Chief amongst them was the Ford Pinto must weigh less than 2 000 pounds (907 kg) and cost less than US$2 000. It had to be stylish and appeal to the 'young' demographic. The Ford Pinto that met these design constraints and was launched to the public can be seen below. Its design was considered both contemporary and appealing to the young target market.

Source: https://ccnwordpress.blob.core.windows.net/journal/2019/10/18406951-1973-ford-pinto-std.jpg

We are told by a science encyclopedia (**https://www.encyclopedia.com/ science/encyclopedias-almanacs-transcripts-and-maps/ford-pinto-case**) that it had;

> a short rear-end, perhaps in imitation of the extremely popular Ford Mustang. This limited the engineers' alternatives for fuel tank safety and placement. The tank was placed behind the rear axle instead of an over-the-axle placement.

Prototypes were made of the Ford Pinto, which underwent crash-testing as required by the National Highway Traffic Safety Administration (NHTSA) to see if it met US safety standards. Whilst the Ford Pinto met the requirements of these tests done in the early 1970s, Ford US were aware the NHTSA were going to significantly change their safety standards in 1972, specifically in relation to a vehicle's ability to withstand rear-end collisions. It was known that in 1972 the NHTSA were going to require that vehicles were able to withstand a rear-end collision of 20 miles per hour (mph) (32 kph) without fuel loss that in 1973 would be increased to 30 mph (48 kph). Ford US were aware of this when they undertook their own in-house tests on their Ford Pinto prototypes and found the Pinto failed the 20 mph rear-end collision test as the fuel tank ruptured and the resulting fuel leaks could represent a fire hazard. This, it was felt, was due to the position of the fuel tank, which was located behind the rear axle rather than above it to create more boot space. In addition, the fuel tank and the rear axle were separated by a mere 9 inches (22.86 cm) and the bolts were positioned in a manner that threatened the fuel tank of the vehicle if a rear-end collision occurred. Furthermore, the fuel filler pipe design resulted in a higher probability that it could disconnect from the fuel tank in the event of a rear-end collision that could in turn cause a fuel spillage that could lead to a fire hazard and compromise the safety of both driver and passengers.

FORD PINTO'S COST–BENEFIT ANALYSIS

The Ford US assembly line was scheduled to start production of the Ford Pinto in August 1970. By the time the Ford Pinto in-house test results were known, the assembly line had already been tooled up. Ford US faced a decision. Should they launch the existing Ford Pinto, knowing that it had serious safety concerns, although it had passed the government NHTSA required tests, or should they make a safety improvement to the fuel tank design at a cost of US$11 per vehicle?

To assist their final decision-making, Ford US used a strict cost–benefit analysis and calculated the cost per projected fatality against the cost of making the changes to the Ford Pinto that would ensure the safety of drivers and passengers in the event of a fire in the Ford Pinto, should the vehicle suffer a rear-end collision of 20 mph or more.

The number of fatalities from rear-end collisions, according to NHTSA data, was 180 people annually, whilst the NHTSA provided an estimate of US$200 725 that society lost every time a person was killed in an auto accident (for itemised costs see Table 16.1).

Table 16.1: The cost per person for an auto rear-end collision fatality as designated by NHTSA in 1971.

Direct loss	$132 000
Indirect loss	$41 300
Medical costs hospital	$700
Medical costs other	$425
Property damage	$1 500
Insurance administration	$4 700
Legal and court expenses	$3 000
Employer losses	$1 000
Victim's pain and suffering	$10 000
Funeral	$900
Assets (lost consumption)	$5 000
Miscellaneous accident costs	$200
TOTAL PER FATALITY	**$200 725**

With these costs in hand, Ford accountants calculated the costs to Ford US of deaths, injuries and the loss of a Ford Pinto vehicle in the benefit statement (shown in Table 16.2).

FORD PINTO'S BENEFITS

Table 16.2: Ford benefit statement.

Savings	180 burn deaths; 180 serious burn injuries; 2 100 burnt vehicles
Unit cost	$200 000 per death; $67 000 per injury; $700 per vehicle
TOTAL BENEFIT	**(180 × $200 000) + (180 × $67 000) + (2 100 × $700) = $49.5 million**

FORD PINTO'S COSTS

Ford US now needed to calculate the costs of making the Ford Pinto 'safe' on their existing assembly line. They calculated their costs as shown in Table 16.3.

Table 16.3: Ford cost statement.

Sales	11 million cars; 1.5 million light trucks
Unit costs	$11 per car; $11 per truck
TOTAL COST	**12.5 × $11 = $137.5 million**

From these statements it can be seen the costs of the suggested safety modifications outweigh their benefits and as such the management recommended against making the safety improvements.

However, not all costs are calculable. Apart from the unethical exercise of Ford US management using a strict cost–benefit approach to calculate the *'value'* of a Ford Pinto consumer versus the *'cost'* of a Ford Pinto consumer to Ford, the accountants **did not calculate** the costs that would be incurred by Ford US if they decided to launch the Ford Pinto and its shortcomings were discovered.

Costs that were not calculated or included in the cost–benefit analysis could be argued to be:

- The costs of negative publicity.

- Legal costs, for example advocates, lawyers, expert witnesses if civil/criminal cases were pursued.

- Costs of settlements/awards and court judgements.

- The negative impact on Ford share prices.

- The cost of a public relations campaign to restore the Ford brand in the eyes of the consumer.

- The loss of the cost to the company and automotive industry of experienced senior management and designers who may be forced to resign their corporate positions.

THE FORD PINTO FIRES: 1971–1978

Between 1971 and 1978 there were a considerable number of deaths and injuries caused by fires from the impact of rear-end collisions to the Ford Pinto in the US. The safety issues became apparent as approximately 50 lawsuits were brought against Ford by both burn victims and the families of those who were fatally injured in the fires in their Ford Pintos. In most cases the victims were young and whilst Ford's lawyers reported there were 23 deaths, it has been claimed that as many as 500 deaths can be attributed to the fires caused by rear-end collisions in the Ford Pinto, although this has never been verified.

It was arguably the Grimshaw case in 1972 and the Ulrich case in 1978, in which three teenagers were burnt to death in their Ford Pinto, that brought Ford's negligence into the spotlight and their reliance on a cost–benefit analysis rather than adhering to their duty as engineers to public health and safety. However, Ford US has always advised the public and the juries that sat on the cases brought against them that its Ford Pinto vehicle met and surpassed the government's own safety standards.

Ford stopped producing the Ford Pinto in 1980 (**https://www.encyclopedia. com/science/encyclopedias-almanacs-transcripts-and-maps/ford-pinto-case**). In the period since its launch the company was estimated to have sold almost three million Ford Pinto vehicles.

LEARNING OUTCOMES

- Explain how a cost–benefit approach is compatible with the Utilitarian approach.

- List the ethical approaches (other than Utilitarianism) that could be applied to the Ford Pinto case study.

- Discuss how Ford chose to ignore both human dignity and human rights for those who were drivers or passengers of the Ford Pinto.

- Were the NHTSA safety standards sufficient for non-commercial vehicles? Explain your position.

- Explain whether an accurate cost–benefit analysis can be undertaken. Are all costs known and calculable?

GRAYSTON BRIDGE COLLAPSE AND TONGAAT MALL COLLAPSE

The following two case studies demonstrate how engineering disasters occur when the duties of a professional engineer are not met and when designs are not adequately examined, assessed and 'signed off' as being both acceptable and compliant by an accredited professional engineer. Both the Grayston Bridge Collapse and the Tongaat Mall Collapse case studies speak to a lack of engineering professionalism and compliance with accepted safety standards. These case studies are also evidence of inadequate risk and safety mitigation strategies being undertaken by engineers, including a lack of oversight and as a result can be seen to encompass a lack of both intellectual and moral virtues by the main engineering contractors.

CASE STUDY: GRAYSTON BRIDGE COLLAPSE

Source: https://www.engineeringnews.co.za/article/m1-bridge-collapse-inquiry-to-be-public-2015-12-09/rep_id:4136

The collapse of the Grayston Bridge (**https://www.news24.com/ citypress/business/damning-report-into-m1-highway-bridge-collapse-that-killed-two-20191202**), which at the time was a temporary structure that was a cantilever spar cable-stayed bridge, occurred on the afternoon of 14 October 2015. One hundred and twenty tons of steel crashed onto the M1, killing two people (a local taxi driver and a Natal businessman) and injuring 19 more. The Grayston Bridge was designed to be a pedestrian and cyclist bridge spanning the M1 (a six-lane highway, three lanes in each direction) in Sandton (Northern Johannesburg). It was constructed parallel to Grayston Drive to accommodate the 10 000 pedestrians who needed to cross the road every day from the Alexandra township for work in Sandton and beyond.

THE DETAILS OF THE GRAYSTON BRIDGE COLLAPSE

The Johannesburg Development Agency (JDA) appointed Murray & Roberts Construction (MRC) as the main contractor. MRC in turn subcontracted the supply of the scaffolding to Form-Scaff. The bridge building project was overseen by Neami Consulting, who were appointed by the JDA to ensure compliance with all safety requirements at the site. Neami's representative tasked with assessing safety compliance for the build of the Grayston Bridge was Roxana le Roux, who, while recording some of the safety deficiencies and non-compliance issues with the bridge build, nevertheless failed to put a stop to engineering work until the problems were corrected and accepted safety standards and practices had been complied with.

Hein Pretorius was appointed contract manager for the Grayston Bridge project by MRC. He was not registered with ECSA and had no experience in bridge building or erecting temporary structures. Oliver Aadnesgaard, a candidate engineer with ECSA, was appointed site engineer despite his inexperience. As such, he should have been carrying out engineering work for the MRC Grayston Bridge engineering project under the supervision of a registered engineer.

It was also reported by *City Press* (**https://www.news24.com/citypress/ business/damning-report-into-m1-highway-bridge-collapse-that- killed-two-20191202**) that, contrary to the terms of the contract signed between JDA and MRC, the latter failed to appoint an engineer specifically to design the temporary structure for the Grayston Bridge that subsequently collapsed. This engineer was also, according to the terms of the contract, required to supervise the erection of the temporary structure and, through regular site inspections, ensure the temporary structure under construction met all the design requirements.

The lack of this appointment resulted in the drawings that were shared between MRC and Form-Scaff for the temporary structure being undertaken purely as a means of determining the number of components needed for the temporary structure and to form the basis of a quote for Form-Scaff to present to MRC for the supply of the required scaffolding for the Grayston Bridge engineering and construction project.

Whilst the Grayston Bridge Collapse was alleged to have been caused by an excessively strong gust of wind, during the subsequent inquiry that was held by the Department of Labour, MRC's expert witness Professor Mostert (**https://ewn.co.za/2016/07/08/Experts-in-Murray-and-Roberts-probe- says-couplers-holding-structures-were-pulled-apart**) gave evidence that proposed the couplers, provided by Form-Scaff and used to hold together the scaffolding's metal structures, did not comply with required standards in South Africa.

However, Form-Scaff provided evidence to suggest the couplers were not tightened sufficiently by those tasked with erecting the temporary structure by MRC. This was further endorsed by evidence confirming there were missing bolts in the scaffolding and girder batteries were misaligned in the temporary structure. It was reported by *City Press* (**https://www.news24. com/citypress/business/damning-report-into-m1-highway-bridge- collapse-that-killed-two-20191202**) the temporary structure over the MI highway was;

kept together by luck rather than engineering skill

and the collapse was caused through engineering incompetence and negligence.

A further cause for concern was that traffic was allowed to flow beneath the bridge, despite there being safety concerns. Once the bridge collapse occurred, the *Sandton Chronicle* (**https://sandtonchronicle.co.za/131345/bridge-collapse-on-m1-highway/**) reported the traffic congestion prevented emergency services and personnel, who were critical to the rescue operations, from arriving at the rescue scene and treating the injured timeously. It was reported that two people remained trapped in the wreckage almost four hours after the collapse.

In the picture below we can see one of the vehicles that was crushed by the bridge collapse and can only imagine the terror of those who were driving along the M1 and unable to avoid the consequences of the scaffolding falling directly onto the highway and the vehicles travelling along it.

Source: https://www.news24.com/citypress/business/damning-report-into-m1-high-way-bridge-collapse-that-killed-two-20191202.

THE CONTRIBUTING FACTORS IN THE GRAYSTON BRIDGE COLLAPSE PRESENTED IN THE INQUIRY

The inquiry into the Grayston Bridge collapse was undertaken by the Department of Labour and the proceedings confirmed a litany of errors that contributed to this engineering disaster. The first inquiry was established in February 2016, resumed in 2017 and held again in 2018. Essentially the evidence presented at the inquiries could be seen to fall into the two principal areas of structural issues and design issues, which in both cases also suffered from poor safety and risk mitigation strategies and were directly impacted by extrinsic forces such as the lack of a positive working relationship between MRC and Form-Scaff, which seemed to suffer from poor communication and trust issues. Furthermore, from evidence provided to the inquiry, the Grayston Bridge collapse could also be attributed to the lack of a proper planned sequence for the construction of the bridge by MRC, who seemed to adopt a rather ad hoc approach to the construction project.

The image below not only provides a rough graphic of the bridge, but more importantly identifies the contributing errors in engineering construction raised in the hearings and where they were found to have occurred in the Grayston Bridge construction, both in its temporary and permanent structures.

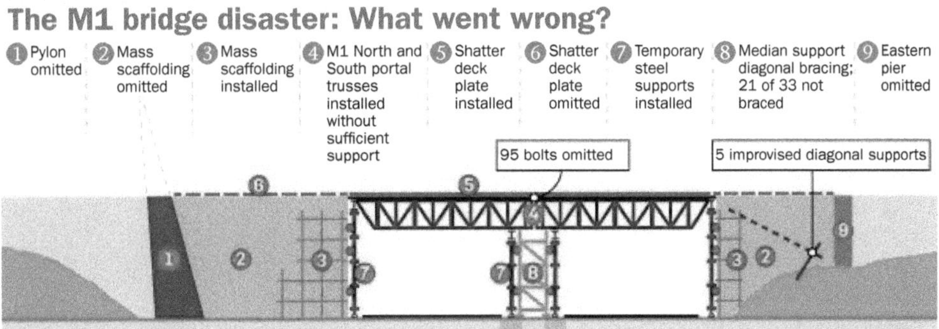

Figure 17.1: The M1 bridge disaster: What went wrong?
Source: https://www.news24.com/citypress/business/damning-report-into-m1-highway-bridge-collapse-that-killed-two-20191202

LEARNING OUTCOMES

- Was the ECSA code breeched by MRC? Discuss.

- List the extrinsic forces that directly impacted the outcomes of this case study.

- Explain how the categorical imperative of an engineer was neglected.

- Were adequate safety and risk mitigation strategies undertaken by MRC?

- Can this case study be approached and understood using a rights approach?

CASE STUDY: THE TONGAAT MALL COLLAPSE

On 19 November 2013 a section of the Tongaat Mall, which was under construction, collapsed after the neck of a column exploded, causing the concrete slab it was supporting to collapse, crushing two workers to death and seriously injuring a further 29 construction workers in what has been described as the one of the worst construction engineering disasters in the KwaZulu-Natal province.

Those living close to the construction site shared with the *Mail & Guardian* newspaper **(https://mg.co.za/article/2013-11-19-one-dead-29-injured-in-tongaat-mall-roof-collapse/)** that since the previous day, scaffolding had started to be removed from the supporting level when late in the afternoon of 19 November 2013, the construction collapsed, trapping construction workers beneath it, who cried out for help whilst being crushed beneath the concrete.

The picture below graphically illustrates the devastation on site and also serves as a reminder of the difficulty experienced by rescue workers, who worked tirelessly to try and free the dead and injured construction workers trapped under the enormous weight of the concrete slab.

Figure 17.2: Devastation on site.
Source: https://mg.co.za/article/2013-11-19-one-dead-29-injured-in-tongaat-mall-roof-collapse/

The disaster formed the basis of a public inquiry in which contraventions of the Occupational Health and Safety Act (OHSA) were found to have occurred in addition to non-compliance with construction regulations. It was reported by News24 (**https://www.news24.com/News24/tongaat-mall-collapse-report-to-be-given-to-npa-20160524**) the Labour Minister at the time, Mildred Oliphant, said the findings from the inquiry were to be handed to the National Prosecuting Authority (NPA) for a decision on whether to prosecute anyone who was directly involved in the engineering construction project. News24 also reported that she further stated;

> the department would not take action against those implicated unless the NPA decided to prosecute.

FACTORS UNCOVERED LEADING TO THE TONGAAT MALL COLLAPSE

At the time of its construction, the Tongaat Mall project was worth R208 million. As investigations were undertaken to try and establish the key factors that led to the Tongaat Mall collapse, it became apparent that several companies were involved in the project, often through the practice of nepotism. In addition, the integrity of the project was arguably further compromised by Jay Singh, the owner of the construction company Gralio Precast, who was a well-known businessman in the area and a significant benefactor of the ruling African National Congress (ANC) party.

Rectangle Property (a property investment company) owned by Jay Singh's son, Ravi Jagadasan, bought the stand on 263 Gopallala Street in Tongaat in 2012 from the Strathmore Property Investment Trust. Included in the purchase price were unapproved plans for a shopping mall that were first submitted to the eThekwini municipality for approval in January 2007. Axion Consulting Engineers Services was appointed as design engineer when the project was first planned by Strathmore and was later appointed in this capacity by Gralio Precast. The latter was engaged by Rectangle Property as an agent and principal contractor to perform all the construction work on the proposed shopping mall.

In February 2013 an earthworks application was submitted to the municipality by Gralio Precast that was subsequently rejected on 20 March 2013. However, excavation work had already begun by Gralio Precast without municipal approval and was at an advanced stage. Moreover, without an approved earthworks application, building plans could not be approved by the municipality, a fact Gralio Precast chose to ignore.

FINDINGS FROM THE INQUIRY

The findings from the inquiry were varied and numerous, and spoke to construction defects, design defects, lack of engineering competency, the quality of the materials used in the construction, and contraventions of the National Building Regulations (NBR) and the Occupational Health and Safety Act (OHSA). All three companies involved in the Tongaat Mall engineering project, ie Gralio Precast, Axiom Consulting Engineers and Rectangle Property, were found to be negligent and non-compliant with accepted standards and practices in South Africa, in addition to not discharging the duties expected from practising engineers.

The inquiry noted the concrete pourings had not been properly undertaken, which resulted in the poor construction of Beam 7, which triggered the collapse. It was typical, the inquiry found, of the poor and often inadequate construction methods that had been found to exist in the Tongaat Mall construction project.

The inquiry also found the piles for some of the columns had been overloaded and under-designed, in addition to a failure of Gralio Precast to work from designs. Instead, changes to designs for the Tongaat Mall were made by Gralio Precast on site, without being professionally drawn up and sufficient load-bearing calculations being made. The inquiry also found there was a lack of knowledge and engineering expertise at Gralio Precast to enable the execution of such a complex and independent structure as presented by the Tongaat Mall construction project.

The inquiry further found an alarming lack of supervision for the construction work. Neither an in-house nor independent supervisor was employed to ensure the quality of the construction work. The inquiry concluded that, as a result, neither construction work nor the quality of the materials being used in the project were checked or supervised. This resulted in defective materials being used in the construction process. This was the case when it was found that cement imported from Pakistan, which had been sourced by the contractor to cut costs and enhance company profits for the construction project, did not comply with the South African Bureau of Standards (SABS) requirements.

THE AFTERMATH

The NPA, who were handed the results from the commission of inquiry into the Tongaat Mall collapse in 2015, did not pursue the case further. As a result, no fines were issued or paid by any of the companies associated with the Tongaat Mall collapse. Mildred Oliphant announced amendments to the OHSA which would see changes to the construction regulations (as proposed by the inquiry in 2014). This would make it significantly more difficult for construction companies to break the law. No compensation has been received by any of the victims of the collapse or their families, whilst Gralio Precast has received a further R79 million worth of construction contracts in 2016 from the ANC-controlled eThekwini Municipality.

LEARNING OUTCOMES

- Did Gralio Precast meet their engineering categorical imperative of putting public health and safety first? Discuss.

- Were sufficient safety and risk mitigation strategies undertaken and enforced by engineering contractors working on the Tongaat Mall project?

- List the areas in which there was an absence of engineering professionalism in this engineering project.

- As an engineering ethicist, discuss how you believe the concept and reality of human dignity and human rights was denied in this case study.

- Explain how and where governance and compliance were ignored in this case study.

COALBROOK MINING DISASTER AND THE BOEING 737 MAX 8

CHAPTER 18

The following two engineering case studies demonstrate how disasters occur when engineering duty is neglected and there is an absence of safety and risk mitigation strategies. Moreover, when examined closely, these two engineering case studies clearly establish how engineering ethics have been overlooked or ignored along with a lack of human dignity and human rights in favour of corporate profits. In both case studies engineering ethics have been neglected, particularly when viewed in light of the engineering methods and practices that have been endorsed by the management in each case study.

CASE STUDY: COALBROOK MINING DISASTER

The Coalbrook Mining Disaster (**https://www.sahistory.org.za/dated-event/more-400-miners-killed-what-was-worst-mining-disaster-south-africa**) that occurred at the Clydesdale Coal Colliery in Coalbrook, near Sasolburg, remains the worst in South African mining history. It happened on 21 January 1960 at around 19:00, when approximately 900 pillars that were almost 180 m underground gave way in what is known as a *'cascading pillar failure'*. This occurs when a few pillars in a tunnel fail, which in turn increases the load on adjacent pillars that cannot take the increased load and fail also, as evidenced in their collapse. It has been reported the pillar collapse in Coalbrook covered an area of 324 ha. About 1 000 miners were in the mine at the time of the collapse and, whilst some of this number managed to escape through an incline shaft, unfortunately 435 miners were trapped by rockfall, whilst others were choked by the methane gas and carbon monoxide that tore through the tunnel they were mining. The number of miners who had been crushed by rocks or suffocated by methane gas and/or carbon monoxide was unknown, as was the number of possible survivors.

A rescue operation was immediately undertaken using a specialist drill from Texas to extract survivors and the bodies of those who had died. However, it could not drill through the dolerite (hard rock) that was above the shaft and the drill bits wore down.

In an effort to continue rescue operations and in the hope of finding survivors, boreholes were subsequently drilled by the rescuers into areas where it was thought survivors might be. Microphones were lowered into these holes to check if there were any signs of life. However, only the sound of water was heard and nothing else. After eleven days the rescue operation was called off and concrete was poured into the rescue shaft, sealing the bodies of all the miners who were underground in the tunnel forever.

Of those miners killed, six were white and the balance (429) were Lesotho and Mozambique nationals. Owing to the policy of apartheid that prevailed at the time, the Workmen's Compensation Act entitled 'white' widows to their deceased husbands' pension funds, whilst 'black' widows were granted a small lump sum from the mining company.

HOW DID IT HAPPEN?

The opening of the coal-powered Taaibos power station in 1954 and the Highveld power station in 1959 created an increased demand for coal and the design of their boilers allowed for a poorer quality of coal to be used than had been the case in other power stations. This provided Coalbrook with the opportunity to extend the mining life of the mine by extracting coal from its shafts – that had previously been uneconomical to mine – using what is called the *'top coaling'* method. This method used pillars to raise the roof height of tunnels in mines. The heights could be raised anything from 4.3 m to 6.1 m (see Figure 18.1, which provides an indication of the coal seams in Coalbrook and their dimensions).

Geology and Dimensions

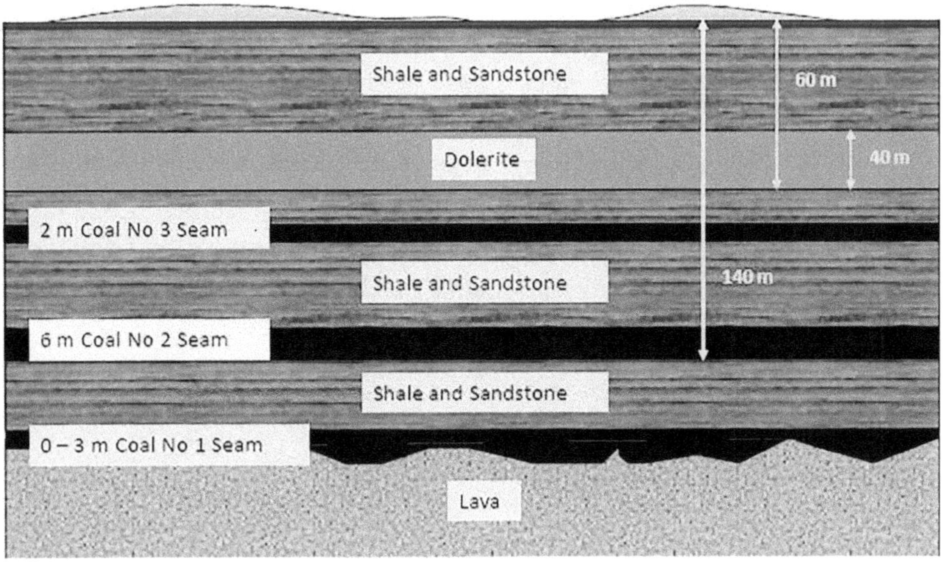

Figure 18.1: Coal seams in Coalbrook Mine and the use of the top coaling method for coal extraction.

Source: https://www.theheritageportal.co.za/article/1960-coalbrook-disaster

Coalbrook experimented with the use of this 'top coaling' method using pillars and raised these tunnels at Section 10 of the mine to allow extraction of the poorer quality coal by miners next to the dolerite rock. However, in adopting this method of mining, a paper in the South African Institute of Mining and Metallurgy (SAIMM) journal by JN van der Merwe (**http://www.saimm.co.za/Journal/v106n12p857.pdf**) suggests;

> Coalbrook had to resort to mining methods with unknown consequences in order to meet the demand for coal from the power station.

A collapse did occur in Section 10 of the mine, in late December 1959, but went unreported. This was in contradiction of the mine safety legislation in place for the South African mining industry, that required the reporting of any mining collapse to the mine inspectors. Instead, the collapse was left unreported by Coalbrook as no fatalities or injuries resulted. Despite the reluctance of some miners to return to their duties after the collapse, all miners were required by mine management at Coalbrook to continue their activities at the mine.

Van der Merwe suggested the investigators into the Coalbrook disaster were left asking questions concerning certain aspects of the 'top coaling' method that had used pillars that subsequently collapsed. Chief amongst these was the strength and load-bearing ability of the pillars erected and used in the tunnels. The questions posed focused on the following aspects:

* The strength of the square coal pillars

* The strength of the barrier pillars

* The strength of the overburden

* The nature of the pillar loading

Interestingly, since the disaster, the only area of mining engineering that is now sufficiently understood by mining engineers is pillar strength. Very little progress has been made in understanding the barrier pillar strength, the overburden or indeed the pillar loading conditions despite the establishment of the Chamber of Mines Research Organisation (COMRO) and the Coal Mines Research Controlling Council (CMRCC), both of which were tasked to conduct research into 'hard rock' mining after the Coalbrook disaster. However, funding has been minimal and this has negatively impacted the amount of research that has been able to be undertaken by mining experts and mining researchers.

Following the Coalbrook disaster, at least four different inquiries were launched under the Mines and Works Act of 1956, but it was found the deaths of the miners occurred as a result of subsidence of the mine itself, a finding that created significant speculation both at the time and subsequently.

LEARNING OUTCOMES

- Should the engineers at Coalbrook have used engineering methods and practices that were in large part experimental? Discuss.

- Were adequate safety and risk mitigation strategies undertaken by the engineers and management of Coalbrook? Discuss.

- Were human dignity and human rights afforded the miners at Coalbrook, both those who died and those who worked in the tunnels?

- Could it be argued that a strict cost–benefit analysis was used to justify the 'top coaling' method at Coalbrook? Explain.

- Was the engineer's categorical imperative of public health and safety exercised at Coalbrook? Discuss.

 CASE STUDY: THE BOEING 737 MAX 8

Source: https://www.theguardian.com/business/2021/nov/11/boeing-full-
responsibility-737-max-plane-crash-ethiopia-compensation

Central to the Boeing 737 Max 8 aircraft aeronautical engineering case study is the Manoeuvring Characteristics Augmentation System (MCAS), a flight-stabilising software program that was designed, developed and installed for the 737 Max 8. This aircraft was launched as Boeing's response to the Airbus A320neo, which had achieved highly successful sales for Airbus SAS in the passenger aircraft market. Airbus SAS is the second largest maker of commercial aircraft globally after Boeing and is considered their main competitor.

ABC News (**https://abcnews.go.com/US/boeing-737-maxs-flawed-flight-control-system-led/story?id=74321424**) reported that when Boeing launched the 737 Max 8 it was;

 the fastest selling jet in the company's history.

The Boeing 737 Max 8 had been designed to produce added power and fuel efficiency to Boeing's passenger aircraft and was essentially a revamp of Boeing's highly popular 737 aircraft. The Boeing 737 Max 8 was also equipped with the newly designed MCAS that software engineers claimed was developed by Boeing to overcome a basic design fault in the 737 Max 8.

Gregory Travis **(https://www.fierceelectronics.com/electronics/killer-software-4-lessons-from-deadly-737-max-crashes)**, a veteran instrument-rated pilot and career software engineer said in an interview that:

> Instead of going back to the drawing board and getting the airframe hardware right, Boeing relied on MCAS. Boeing's solution to its hardware problem was software.

The 737 Max 8 took its maiden flight in early 2016 and was certified by the United States Federal Aviation Administration (FAA) in March 2017. The first delivery of the Boeing 737 Max 8 aircraft took place in March 2017.

The first fatal aircraft crash of a Boeing 737 Max 8 took place on 29 October 2018 when Lion Air Flight 610 crashed into the Java Sea 13 minutes after take-off from Soekarno-Hatta International Airport, killing 189 passengers and crew. The second fatal aircraft accident involving a Boeing 737 Max 8 took place on 10 March 2019 when Ethiopian Airlines Flight 302 crashed six minutes after take-off from the international airport in Addis Ababa. The plane penetrated almost 14 m into the ground on impact and 149 passengers and 8 crew on board were killed.

In subsequent investigations the MCAS developed and installed in the 737 Max 8 by Boeing was found to be the cause of these two fatal aircraft accidents that claimed the lives of 346 passengers and crew on both flights.

Boeing had 378 of the 737 Max 8 aircraft in service around the world at the time of the two fatal aircraft crashes.

FIRST IN-FLIGHT SUSPICIONS OF THE BOEING 737 MAX 8 AND THE MCAS

Lion Air (**https://www.nytimes.com/2019/09/18/magazine/boeing-737-max-crashes.html**) was one of Boeing's largest buyers of the 737 Max 8 passenger planes, ordering over 200 at a total cost of $22 billion. Its first flight using this aircraft was in May 2017 and was without incident, as were all commercial flights for 17 months. However, on 28 October 2018, Lion Air Flight 610 from Bali to Jakarta experienced an in-flight emergency. This was arguably the first time on a commercial flight that it was reported a 737 Max 8 suddenly nosedived after take-off. It was reported by the pilots the aircraft nosedived four times as they struggled to regain control of the malfunction. It was only due to the efforts and interventions of a third pilot, who was able to resolve the situation and gain control of the plane through his knowledge of shutting off the *'electric trim'* by flipping the cut-out switch that in turn disabled the MCAS, that resolved the problem. His prompt actions averted a crash and the plane was subsequently landed safely in Jakarta by its pilots.

Whilst the pilots reported the problems experienced after take-off, ABC News (**https://abcnews.go.com/US/boeing-737-maxs-flawed-flight-control-system-led/story?id=74321424**) claimed that a senior safety investigator from Indonesia's Transportation Safety Committee (NTSC) said the pilots involved in this incident left incomplete notes and records that simply highlighted:

[T]hey had a problem with the speed and altitude indicated on the captain's side of the cockpit.

The senior investigator said the captain had failed to mention the aircraft's *'trim system'* (MCAS) had inexplicably activated, causing it to repeatedly nosedive. Early the next day (19 October 2018) the same plane departed from Jakarta (Lion Air Flight 610) and crashed 13 minutes after take-off into the Java Sea. The flight data recorder recovered from the crash revealed the plane had gone out of control and had moved uncontrollably up and down over 24 times before finally crashing into the sea.

ETHIOPIAN AIRLINES FLIGHT 302

The Ethiopian Airlines flight 302 crashed six minutes after take-off, a mere four months after the fatal Lion Air crash in October 2018 (see Figure 18.2). Analysis of the data from the flight recorder revealed similar issues to those discovered in the Lion Air Flight 610 fatal air crash. The Ethiopian Airlines flight recorder data recovered from the crash site pointed to a defective Angle of Attack (AOA) sensor as the cause of the crash. The Boeing 737 Max 8 had two AOA sensors fitted on either side of the fuselage nose, but the flight investigators found that one of them had been transmitting incorrect information about the position of the plane's nose to the MCAS, which was relying on data from just the one sensor. The defective AOA sensor had unfortunately triggered activation of the MCAS, that in turn sent the plane into a nosedive from which the pilots could not recover or regain control.

Figure 18.2: The crash site of Ethiopian Airlines Flight 302.
Source: https://phys.org/news/2019-03-ethiopian-airlines-mcas-boeing-max.html
Photo credit: Michael Tewelde/AFP via Getty Images

Within a week of this accident the *New York Times* (**https://www.nytimes. com/2019/09/18/magazine/boeing-737-max-crashes.html**) reported all Boeing 737 Max 8 planes worldwide were grounded.

BOEING AND REGULATORY FAILINGS

In 2019, the Boeing 737 Max 8 was grounded by global regulators worldwide after a malfunctioning MCAS was identified as being the cause of two fatal aircraft crashes. In redesigning their original 737 aircraft, Boeing made significant design changes to their 737 Max 8.

In moving the wings and the engines forwards in their 737 Max 8 design, Beasley Allen (**https://www.beasleyallen.com/aviation-accidents/boeing-737-max-8-crashes/**) reported that Boeing had;

> changed the aircraft's flight characteristics and caused the nose of the plane to have a tendency to pitch up while in flight. Continued, uncontrolled upward pitch will cause an aircraft to lose forward airspeed and eventually stall.

Faced with this situation, it was suggested by Beasley Allen that:

> Boeing decided to simply patch the deadly flaws with a new automatic flight-control MCAS system.

Boeing wanted the FAA to certify their new 737 Max 8 quickly in order to establish market share and corporate profits and the way in which to achieve this, was to present the aircraft as being just another version of their established Boeing 737 commercial passenger plane. The reason behind this was that it would limit the need for Boeing to require additional training and simulation training to be undertaken by pilots of the 737 Max 8. This represented a considerable cost-saving to their major commercial airline customers who would purchase the 737 Max 8.

Boeing convinced the FAA the addition of the MCAS in their 737 Max 8 ensured their new passenger plane would handle similarly to the earlier versions of their Boeing 737. Boeing never admitted the MCAS was primarily a 'new' stability program providing an anti-stall function for the 737 Max 8, which would have in turn have caused them to incur greater certification requirements and associated costs from the FAA and other aviation authorities around the world. Worryingly, the MCAS also did not feature in the 1 600-page flight manual included in the delivery of every Boeing 737 Max 8, but this was overlooked. Instead, the *New York Times* (**https://www.nytimes.com/2019/09/18/magazine/boeing-737-max-crashes.html**) accused Boeing, in their marketing literature, of confirming their new 737 Max 8 had;

the same pilot type rating, same ground handling, same maintenance program, same flight simulators, same reliability.

After the fatal air crashes, a whistle-blower from within the Boeing corporation advised the FAA did not correctly handle the certification of the 737 Max 8. Beasly Allen claimed that Boeing were aware of the problems with the flawed MCAS, but submitted false data and information about it in self-certifying reports and reviews of their own product to the FAA. This resulted in the Boeing 737 Max 8 being certified by the FAA on 8 March 2017.

Between 2018 and 2019 (just before the fatal air crashes) the FAA received numerous pilot complaints of the *'unexpected behaviour'* of the Boeing 737 Max 8 and how the crew manual lacked any description of the MCAS system, although it appeared in the flight manual glossary. However, no significant action was taken by Boeing or the FAA until the fatal air crashes occurred.

INVESTIGATIONS AND INQUIRIES

After the Lion Air tragedy the CEO of Boeing tried to reassure the public when, in an appearance on Fox Business News on 13 November 2018, he said:

The bottom line is the 737 Max 8 is safe and safety is a core value for us at Boeing.

Unbeknownst to the public, Boeing also simultaneously issued a bulletin to airlines in which they instructed pilots to shut off power to the horizontal stabiliser in the event of an uncommanded nose-down emergency. This would prevent MCAS from activating and subsequently could be argued was a tacit acceptance from the manufacturer Boeing of the failure of their MCAS software.

However, despite the knowledge of the fatal shortcomings of the MCAS, Boeing 737 Max 8 aircraft continued flying until the Ethiopian Airlines Flight 302 fatal air crash. After this and the subsequent findings from the flight recorder, countries around the world grounded all Boeing 737 Max 8 flights. The FAA was one of the last to ground the aircraft.

One hundred victims' families filed a lawsuit against Boeing in addition to fighting to be a part of the congressional investigation that was to take place. The House Committee on Transportation and Infrastructure launched an 18-month investigation into Boeing's development of the 737 Max 8 in order to establish the plane's defects and how the regulatory system failed and allowed the plane to be manufactured and sold with those defects. When the House investigation committee asked who bore the responsibility for the two plane crashes, ABC News **(https://abcnews.go.com/US/boeing-737-maxs-flawed-flight-control-system-led/story?id=74321424)** reported that Dennis Muilenburg, the CEO of Boeing, admitted:

Mr Chairman, my company and I are responsible. We are responsible for our airplanes and we know there are things we need to improve.

In the final report from the House investigation committee, released on 16 September 2020, it concluded that Boeing had knowingly put the safety of the flying public at risk for commercial gain.

The FAA cleared the return to service of the Boeing 737 Max 8 on 18 November 2020 subject to mandated design changes to the MCAS software that ensures it is now under the pilot's control, in addition to mandating ground training and simulator training for pilots scheduled to fly the plane.

The first Boeing 737 Max 8 to fly after the global grounding and the FAA lifting was an American Airlines aircraft that flew on 29 December 2020.

Boeing estimated their order book for the 737 Max 8 had been reduced by in excess of 1 000 aircraft due to cancellations arising from a loss of trust from airlines in the aircraft. In addition, it would appear airlines were noting that passengers who had been aware of the MCAS scandal were not prepared to book flights on a 737 Max 8 even if it was now declared 'safe' by regulatory authorities.

The Guardian (**https://www.theguardian.com/business/2021/nov/11/boeing-full-responsibility-737-max-plane-crash-ethiopia-compensation**) reported that in January 2021 Boeing agreed to a $2.5 billion settlement with the US Department of Justice in fines and compensation. This included a $500 million fund to compensate families of the 346 killed in both 737 Max 8 aircraft crashes.

LEARNING OUTCOMES

- List the extrinsic forces that could be seen to guide Boeing's decision to manufacture and sell an unsafe passenger aircraft.

- Did Boeing undertake adequate safety and risk mitigation strategies for the 737 Max 8? Discuss.

- List the moral and intellectual virtues that were absent in Boeing's management and engineers in their launch and sale of the 737 Max 8.

- Explain how Boeing denied passengers and crew on the 737 Max 8 their human dignity and human rights.

- In your opinion, should regulatory authorities be subject to additional oversight in an attempt to restore their ethical and moral conduct?

WHISTLE-BLOWING:

GAUTENG DEPARTMENT OF HEALTH AND PASSENGER RAIL AGENCY OF SOUTH AFRICA (PRASA)

Whistle-blowing has been covered quite extensively in an earlier section of this book, in addition to the five criteria or conditions that need to be met for successful whistle-blowing. There is consensus that whistle-blowing is never easy and definitely not for those who have any personal doubts about their dedication and determination when trying to address unethical and/or illegal activity and behaviour.

Whistle-blowing, especially when it is acknowledged, can be accompanied by significantly adverse consequences for the whistle-blowers. The fourth condition of 'last resort' for successful whistle-blowing and the fifth condition of 'personal commitment' in successful whistle-blowing often bring with them personal sacrifice and professional hardships for the whistle-blower(s). This is often referred to when the US engineering case studies of Goodearl and Aldred versus Hughes Aircraft Corporation and the Bay Area Rapid Transport Company (BART) are examined. In both case studies the whistle-blowers experienced personal and professional hardships because of their whistle-blowing and what some would refer to as their refined sense of ethics and morals. Interestingly, when interviewed, the whistle-blowers concurred, they had *done the right thing* and, faced with the same set of circumstances again, would have undertaken the same course of action.

Recently in South Africa, the fifth condition of 'personal commitment' in successful whistle-blowing has taken on deadly consequences, with the assassination of 53-year-old Babita Deokaran on 23 August 2021 outside her home in southern Johannesburg by six assailants who hailed from KwaZulu-Natal and were described by Global Initiative **(https:// assassination.globalinitiative.net/face/babita-deokaran/)** as;

contract killers willing to murder for a fee.

CASE STUDY: GAUTENG HEALTH DEPARTMENT

Deokaran was a civil servant with 30 years' experience and was the Chief Director of Financial Accounting in the Gauteng Department of Health at the time of her death. She had discovered rampant corruption on a grand scale within the Gauteng provincial government and took the decision to become a whistle-blower to draw attention to the unethical and illegal activities and behaviours that were being undertaken in her workplace, specifically surrounding the awarding of tenders for Covid-19 personal protective equipment (PPE) during the pandemic.

Johann van Loggerenberg (**https://assassination.global-initiative.net/face/babita-deokaran/**), a former executive in the South African Revenue Service (SARS) and himself a whistle-blower to 'State Capture', said:

> When you have a witness in such a matter, aside from extracting the evidence from the witness and using them for testifying in court, as the state, you have a responsibility to take care of their well-being.

This was not the case with Deokaran, whose death, according to Global Initiative (**https://assassination.globalinitiative.net/face/babita-deokaran/**), caused Chief Justice Raymond Zondo, who chaired the Judicial Commission of Inquiry into Allegations of State Capture, Corruption and Fraud in the Public Sector including Organs of State, better known as the Zondo Commission or State Capture Commission, to;

> recommend that protection of whistle-blowers ought to be drastically improved.

He also recommended that an anti-corruption agency be set up with the explicit task of managing whistle-blower disclosures in South Africa. He felt this might also serve to provide the much-needed security for the whistle-blowers, who faced significant personal danger by testifying to unethical and/or illegal practices and behaviours in their workplaces.

With the background of the Deokaran death ever present and the criminal case still pending, the next case study is a South African engineering case study of whistle-blowing in the Passenger Rail Agency of South Africa (PRASA). The whistle-blower is Martha Ngoye, who is worried her actions could lead to her suffering the same fate as Babita Deokaran as she too has been an acknowledged whistle-blower to corruption and State capture in this state-owned enterprise (SOE) which, according to EWN News **(https://ewn.co.za/2022/08/25/prasa-whistle-blower-ngoye-i-fear-i-could-face-same-fate-as-babita-doekaran)**, Chief Justice Raymond Zondo has accused of sliding into *'almost total ruin'*.

Martha Ngoye was PRASA's head of legal and spoke out against what she considered were unethical practices undertaken by senior executives in the SOE, specifically the awarding of the *'Swifambo'* and *'Siyangena'* engineering contracts. She appeared in 2021 at the State Capture Commission, presided over by Chief Justice Raymond Zondo. According to Eyewitness News **(https://ewn.co.za/2022/06/23/many-ills-not-yet-uncovered-zondo-recommends-special-inquiry-into-prasa)**, she was also credited as;

one of the PRASA officials who helped stop the rail agency from investing R1 billion with VBS Mutual Bank, a few months before its collapse.

EWN News (**https://ewn.co.za/2022/08/25/prasa-whistle-blower-ngoye-i-fear-i-could-face-same-fate-as-babita-doekaran**) has reported that, like many whistle-blowers before her and;

despite the lack of protection and past intimidation, Ngoye said she will continue speaking out against corruption at PRASA, because she believes it's the patriotic thing to do.

At the time of appearing at the State Capture Commission in June 2021, Ngoye was suspended from her position as head of legal at PRASA pending an internal disciplinary hearing. PRASA's relationship with Ngoye continued to be judicially adversarial.

CASE STUDY: PASSENGER RAIL AGENCY OF SOUTH AFRICA (PRASA)

The Zondo report on PRASA shared during the inquiry on the State Capture Commission, dealt specifically with the *'Swifambo'* and *'Siyangena'* engineering contracts that arguably served as seminal case studies of corruption and State capture within the SOE within its recent past. These had been the subject of an earlier investigation and report in 2015 called 'Derailed' by Thuli Madonsela (the Public Protector at the time), who had been informed of the maladministration and the impropriety in the awarding of tenders at PRASA. Her report, according to News24 (**https://www. news24.com/Fin24/10-of-thuli-madonselas-findings-against-prasa-20150824**) was a response to 32 complaints laid against the rail agency, of which 19 were substantiated, and which she said provided evidence of;

widespread evidence of maladministration, improper conduct and nepotism.

She also reported in her executive summary of Report No 3 of 2015/16 (**https://www.politicsweb.co.za/documents/how-prasa-was-derailed--thuli-madonsela**) there was clear evidence of;

> victimisation of whistle-blowers by the CEO and other functionaries at PRASA.

Thuli Madonsela's report was the prelude to a forensic investigation into both senior executives of PRASA and their administrative practices, where evidence was found of *'systematic failure'* to comply with the supply chain management policies within PRASA. This led in a number of cases to transport unions bringing complaints against the SOE in 2012, substantiated alongside further evidence provided by Martha Ngoye. These complaints, supported by evidence, led to both Thuli Madonsela's report and subsequently that of Chief Justice Raymond Zondo's State Capture Commission investigation into PRASA.

PRASA

The national government website (**https://nationalgovernment.co.za/units/view/144/passenger-rail-agency-of-south-africa-prasa**) says that PRASA consists of four branches, namely Metrorail, Shosholoza Meyl, Autopax and Intersite, whose activities and management are overseen by the Department of Transport (DOT).

However, even with governmental oversight it would appear that in recent years PRASA has fallen victim to extensive corruption, maladministration and State capture specifically in the cases of the R2.6 billion Swifambo tender for locomotives, which according to News24 (**https://www.news24.com/fin24/Companies/spain-investigates-corrupt-too-tall-trains-sale-while-sa-has-taken-no-action-20220812**) has become known as the *'too-tall trains deal'* and the R6 billion Siyangena Technologies contract that can be viewed as engineering technology related to the PRASA infrastructure.

The PRASA CEO at the time of both cases was Tshepo 'Lucky' Montana (who was appointed to serve a third three-year term at PRASA in 2013). His chief procurement officer was Chris Mbatha.

SIYANGENA TECHNOLOGIES

In October 2020 Groundup (**https://www.groundup.org.za/article/high-court-sets-aside-dodgy-siyangena-contracts/**) reported that three judges sitting in the North Gauteng High Court had set aside all agreements between PRASA and Siyangena Technologies. The latter was reported by Groundup (**https://www.groundup.org.za/media/uploads/documents/siyangena_judgment.pdf**) to have;

unlawfully won billions of rands in contracts to supply security infrastructure to PRASA stations for the 2010 World Cup and subsequent contracts. These items included automated speedstiles, information boards, CCTV, lights and communication systems, to a contract value of approximately R6 billion.

The judges found collusion and corruption between Mario Ferreira, head of Siyangena Technologies, and Lucky Montana, CEO of PRASA at the time. Interestingly, News24 (**https://www.news24.com/fin24/Economy/high-court-sets-aside-r45-billion-prasa-security-contract-with-siyangena-technologies-20201008**) reported that when Siyangena was awarded the initial contract, the budget was R517 million, but after being extended to September 2014 because of the need for further technology support, it *'ballooned'* to R6 billion:

They alleged that Montana had abused his power by awarding tenders to his friends and those loyal to him.

It was further claimed by engineers within PRASA that some of the technologies delivered to PRASA by Siyangena did not constitute leading-edge support technology for the rail agency.

On investigation by the State Capture Commission it was found the contract with Siyangena was subject to irregular and corrupt administrative practices, resulting in personal financial gain for some senior executives at PRASA.

Lucky Montana was released from serving his contractual notice period at PRASA after being implicated in bad practices and had to leave the SOE on 15 July 2015 after the public protector's report was issued. He was called to testify by Chief Justice Raymond Zondo at the State Capture Commission in 2021.

SWIFAMBO 'TOO TALL TRAINS'

Figure 19.1 shows an image of the Afro 4000 diesel locomotives bought by PRASA from Vossloh España who were awarded a tender for 70 of these locomotives for use on South Africa's rail network.

Figure 19.1: The Afro 4000 diesel locomotive.
Source: https://en.wikipedia.org/wiki/South_African_Class_Afro_4000#/media/ File:Class_Afro4000_4007.jpg

The contract placed by PRASA for diesel locomotives was undertaken upon the advice of and with reference to the engineering knowledge of their chief engineer Dr Daniel Mthimkulu.

BusinessTech (**https://businesstech.co.za/news/trending/92734/top-train-engineer-doesnt-need-to-be-registered-prasa/**) reported that PRASA spokesperson Moffet Mofokeng said their chief engineer Dr Daniel Mthimkulu was *'not required to be a registered engineer'*.

ECSA confirmed that Dr Daniel Mtimkulu was not registered with their engineering council despite BusinessTech (**https://businesstech.co.za/news/trending/92734/top-train-engineer-doesnt-need-to-be-registered-prasa/**) reporting that another news agency Netwerk24 had said that;

engineers are required to be registered with ECSA in order to do work for the State.

The reason this was brought to the attention of the South African public was that when PRASA took delivery of the first 13 Afro 4000 diesel locomotives from an original order placed by them for 70 locomotives from the Spanish manufacturer Vossloh España, they were found to be *'too tall'* for local use in South Africa.

Businesstech reported that senior Transnet engineers familiar with the South African rail regulations and requirements had said:

The locomotives have a roof height of 4,264 mm while the maximum height for diesel locomotives may not exceed 3,965 mm.

Businesstech (**https://businesstech.co.za/news/government/92424/south-africas-r600-million-train-blunder/**) reported that PRASA's chief engineer Dr Daniel Mthimkulu appeared in the Johannesburg's magistrate's court in July 2015 on charges of fraud and 'uttering'. The latter in this case referred to the validity of his academic qualifications, which were suspected of being fraudulent. Interestingly, this could perhaps explain why ECSA denied his application for registration to the engineering council in 2006.

Dr Mthimkulu resigned from PRASA in mid-July 2015 just after Lucky Montana's dismissal from the SOE.

LEARNING OUTCOMES

- Describe how this case study supports the requirement for registered engineers to lead engineering activities and projects in South Africa.

- Discuss whether Martha Ngoye has met all five conditions for successful whistle-blowing.

- Dr Daniel Mthimkulu was not a registered engineer with ECSA. If he had been, list the areas of ECSA's ethical code in which he could be accused of being non-compliant.

- Should acknowledged whistle-blowers in South African SOE's be provided with government security for their protection? Discuss.

- Do the results of successful whistle-blowing in South Africa send a clear message to both government and industry of the primacy of ethics for the well-being and flourishing of South African society? Discuss.

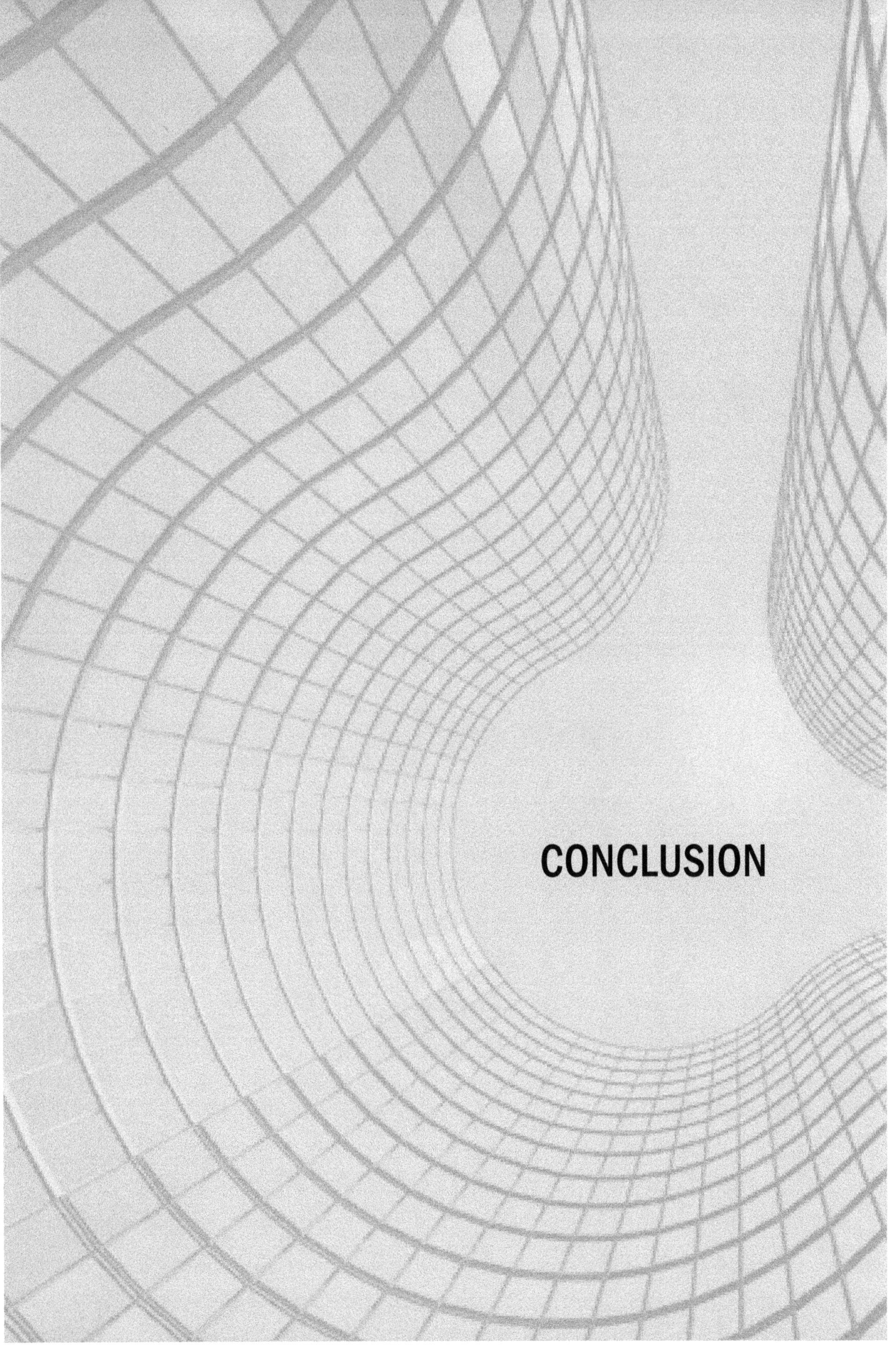

CONCLUSION

WHY DO ENGINEERS NEED ETHICS?

Having read and studied the contents of this book, the reader should now understand there are several responses one could make to the question, 'Why do engineers need ethics?' The immediate answer would be that ethics increase their ability to deal more effectively with the moral complexity in engineering activities, engineering projects and engineering designs and to take this knowledge as professionals into their engineering workplace. However, the immediate answer to the question is rarely *'simple'* as invariably there is a complexity that is hidden from view. Such is also the case with the immediate answer to the question, 'Why do engineers need ethics?'

Increasingly, engineering in southern Africa and around the world is undertaken within a social context that can add to the moral complexity of engineering activities and projects that in turn require a contemporary engineer to develop an *'independent'* sense of moral concern and value. In other words, contemporary engineers as *'independent and autonomous moral agents'* should strengthen their personal moral/ethical abilities, which in turn will develop in them both a sense and practice of what is 'right' and 'wrong', 'good' and 'bad'. This will result in contemporary engineers engaging and acting in ethically appropriate ways to ensure that well-being and flourishing occurs when practising their profession and when working with clients and the community.

The correct understanding and practice of *'moral autonomy'* can provide the means to overcome moral dilemmas and/or conflict and in some instances be the means to prevent their occurrence in the engineering workplace. Such is the case for engineers in South Africa. Moreover, local engineers must be mindful that engineering has its own professional code of conduct that embraces ethics, from which not only public expectations flow, but from which professional virtues and duties flow. This professional engineering code of ethics and conduct can also serve to empower engineers in terms of their moral autonomy because it provides aspects of both governance and compliance within its framework. This serves to outline to both the engineering profession and the public the measurable expectations of how a professional engineer should exercise moral autonomy when engaged in engineering projects and engineering activities.

ETHICAL SKILLS REQUIRED BY A PROFESSIONAL ENGINEER

Martin and Schinzinger (2005) suggest that for engineers to have an overall sense of ethics within their profession, they need to acquire the following ten ethics skills. Table 20.1 identifies these skills and how they need to be demonstrated by engineers working in a social context. After reading through the case studies earlier in the book, it might be a useful exercise for the reader to assess which ethics skills were lacking or absent in the engineers associated with each case study.

Table 20.1: Ten ethics skills required by professional engineers.

SKILL	DEMONSTRATION
1: Moral awareness	Be proficient in recognising and assessing moral dilemmas/conflicts in engineering activities/projects.
2: Cogent moral reasoning	Use the three Cs to assess arguments on opposing sides of moral issues.
3: Moral coherence	Form consistent and comprehensive positions based on relevant facts.
4: Moral imagination	Discern alternative responses/solutions to moral dilemmas/conflicts and create solution(s) to practical difficulties.
5: Moral communication	Practise the three Cs to support moral views adequately to others.

SKILL	DEMONSTRATION
6: Moral reasonableness	Be willing and able to be morally reasonable and seek consensus when required.
7: Respect for persons	Show concern for the well-being of others and oneself and be ever-mindful of the concept and practice of human dignity.
8: Tolerance of diversity	Have respect for ethnic and religious differences and accept reasonable differences in moral perspectives.
9: Moral hope	Have an appreciation for the possibilities of the use of rational dialogue in resolving moral dilemmas/conflicts.
10: Moral integrity	Maintain moral integrity, particularly through seeking and practising consistency in personal and professional ethics.

In acquiring these ten ethics skills, it is hoped that ethical decision-making by engineers in South Africa will be guided to ensure that well-being and flourishing will be at the epicentre of all engineering activities and projects undertaken by them. In so doing, engineers in South Africa must also honour their professional codes when working in a social context to ensure that professional duties and expectations of both ECSA and the public are met.

References

Martin, MW & Schinzinger, R (2005) *Ethics in Engineering*. Fourth Edition. US: McGraw-Hill.

ACKNOWLEDGEMENTS

Rhino poaching – Page 12 – Via Getty Images

Xenophobia in South Africa – Page 13 – Photo by Gallo Images / Sunday Times / James Oatwa

ECSA's Code of Conduct – Page 107 – Engineering Council of SA in *Government Gazette* 17 March 2017

Risk management process – Page 123 – Disclaimer used. Oehmen, J., Ben-Daya, M., Seering, W., Al-Salamah, M.: Risk Management in Product Design: Current State, Conceptual Model and Future Research. DETC2010-28539. Proceedings of the ASME 2010 International Design Engineering Technical Conference & Computers and Information in Engineering Conference IDETC/CIE 2010. August 15-18, 2010, Montreal, Canada. ISBN 978-0-7918-3881-5

The Kariba Dam wall under construction – Page 148 – http://www.zimbabweconnections.com/lake-kariba/ Used by permission of Sean Kelly MBA Zimbabwe Connections Safaris

The completed dam wall – Page 147 – Photo by DeAgostini/Getty Images

How exposed to danger the construction workers were on this project – Page 107 – Terrence Spencer 1955. This work was first published in Zimbabwe (or one of its antecedents) and is now in the public domain in Zimbabwe because its copyright protection has expired by virtue of the Copyright and Neighboring Rights Act, enacted in 2000

A graphic for the semi-dry paddock method and illustrates the water gain and loss in a typical tailings dam – Page 152 – Wills' Mineral Processing Technology 2016. Author: Barry A. Wills, James A. Finch © and permission of Elsevier, Science Direct

Ford Kuga SUVs that have spontaneously caught alight in South Africa © and permission of Warren Krog – Page 166 – Owner of the burning car, who took the photo

Ford Kuga SUVs that have spontaneously caught alight in South Africa – Page 166 – Disclaimer used.

The Ford Pinto – Page 171 – Disclaimer used. Photograph by Shane McGlaun

Grayston Bridge Collapse – REUTERS/Siphiwe – Page 177 – Copyright © thomsonreuters.com. Sibeko via GALLO IMAGES

One of the vehicles that was crushed by the bridge collapse – Page 180 – Disclaimer used. Photo by Dan Maswanganye (https://www.news24.com/citypress/business/damning-report-into-m1-highway-bridge-collapse-that-killed-two-20191202)

The M1 bridge disaster: What went wrong? – Page 181 – Damning report into M1 highway bridge collapse that killed two. (Text by Gallo Images/ Rapport)

Devastation on site – Page 183 – Disclaimer used. SAPA closed in 2015

Coal seams in Coalbrook Mine and the use of the top coaling method for coal extraction – Page 189 – Disclaimer used

THE BOEING 737 Max 8 REUTERS/Karen Ducey – Page 192 – TPX IMAGES OF THE DAY

The crash site of Ethiopian Airlines Flight 302 – Page 195 – MICHAEL TEWELDE/AFP via Getty Images © 2019 AFP Agence France Presse

The Afro 4000 Diesel locomotives bought by PRASA from Vossloh España Colonel André Kritzinger – Page 207 – The copyright holder of this work, hereby publishes it under the following licence: Creative Commons Attribution-Share Alike 3.0 Unported licence

INDEX

Please note: Page numbers in *italics* refer to figures or tables.